NANOTECHNOLOGY SCIENCE AND TECHNOLOGY

DEVELOPMENT OF NANOTECHNOLOGY IN TEXTILES

NANOTECHNOLOGY SCIENCE AND TECHNOLOGY

Additional books in this series can be found on Nova's website under the Series tab.

Additional E-books in this series can be found on Nova's website under the E-book tab.

MATERIALS SCIENCE AND TECHNOLOGIES

Additional books in this series can be found on Nova's website under the Series tab.

Additional E-books in this series can be found on Nova's website under the E-book tab.

NANOTECHNOLOGY SCIENCE AND TECHNOLOGY

DEVELOPMENT OF NANOTECHNOLOGY IN TEXTILES

A. K. HAGHI
AND
G. E. ZAIKOV
EDITORS

nova publishers
New York

Additional color graphics may be available in the e-book version of this book.

LIBRARY OF CONGRESS CATALOGING-IN-PUBLICATION DATA

Development of nanotechnology in textiles / editors, A.K. Haghi, G.E. Zaikov.
p. cm.
Includes index.
ISBN 978-1-62081-030-9 (soft cover)
1. Nanostructured materials. 2. Textile fabrics. 3. Nanofibers. 4. Nanotechnology. I. Haghi, A. K. II. Zaikov, G. E. (Gennadii Efremovich), 1935-
TA418.9.N35D475 2012
677'.02832--dc23
2012002180

Published by Nova Science Publishers, Inc. + New York

CONTENTS

PREFACE

The textile industry is an early adopter of new ideas and technologies. Textiles are not only for the fashion conscious, they have important applications in the aerospace, automotive, construction, healthcare and sportswear industries. Already on the market are socks and leisurewear with embedded silver nanoparticles that combat odor through killing bacteria – and this capability has been extended successfully to wound dressings.

Across the globe, a tremendous amount of research is taking place in electrospinning techniques. The spun, polymer-based nanofibres can be 'loaded' with different additives which could be nanoparticles, enzymes, drugs or catalysts. Some combinations can be antibacterial and sprayed onto wounds as a kind of healing 'web', others can be conductive or even form filters or membranes.

Scientists are also working on nanoelectronic devices that can be embedded into textiles to provide special support systems for individuals in dangerous professions or sports. Some garments can now provide life-signs monitoring, internal temperature monitoring, chemical sensing and also power generation and storage to enable communication with the outside world. colleagues, or even for skiers or their rescuers to give early warning signs of hypothermia.

Nanotechnology is, broadly speaking, the science of manipulating and rearranging individual atoms and molecules to create useful materials. "Nano" textiles are those designed and engineered on the nanoscale to create specific functions. These nano functions are very diverse and range from UV protection and antibacterial functions to moisture management. The book provides the reader with an overview of the topic.

The subject nanotechnology in textiles is as vast as the universe so to restrict it in a few words is merely impossible. This book summarizes the recent development of nanotechnology in textile areas.

In some establishments, research is ongoing into man-made nanofibers where clay minerals, carbon nanotubes or nanoparticulate metal oxides are used to impart new properties. These properties provide halogen-free, flame retardancy for a fabric, increased strength and shock-absorbency, heat and UV radiation stability, and even brighter coloration.

A.K.Haghi
University of Guilan (Iran)
G.E. Zaikov
Russian Academy of Sciences

In: Development of Nanotechnology in Textiles ISBN: 978-1-62081-030-9
Editors: A. K. Haghi and G. E. Zaikov ©2012 Nova Science Publishers, Inc.

Chapter 1

SILVER NANOPARTICLES PLATING OF FABRICS

M. Saberi Motlagh, V. Mottaghitalab, O. Emamgholipour and A. K. Haghi*

Textile Engineering Dept, Faculty of Engineering,
University of Guilan, Rasht, Iran

1. INTRODUCTION

The coating of metallic particles on polymer surfaces has become an area of interest for many researchers. The coating of polymers using metallic particles in micro and nanoscale extremely enhances the range of their applications for either decorative or superior performance. Regarding this, the applications of textiles as a polymeric substrate are not limited to apparel. This approach provides eye-catching superior functions for textile substrate such as ultraviolet protection, water repellency, electromagnetic interference shielding (EMIS), antistatic behavior and also creates high added value and brilliant decorative effects [1-6].

The common methodology for metal coating consists of the laminating of metal foil, spraying conductive paints and lacquers, sputter coating, flame and arc spraying of metal particles and electroless plating of a variety of metals using chemical reactions.

* Corresponding author: Haghi@Guilan.ac.ir.

Electroless plating was developed to deposit metal from its solution on any kind of substrate without an external current. Metal ions in aqueous solution can be reduced to the metal particles using a reducing agent. Using electroless plating, metallic particles can be deposited even along edges, inside holes, and over rough surfaces which are difficult to be plated [5-10]. Therefore, electroless plating is a preferred way for metal-coating on textiles surface with very high roughness.

In this research, the cotton fabric was metalized by means of silver electroless plating and the final properties of the metalized fabrics were evaluated.

2. Experimental

2.1. Chemical and Reagents

White Cotton fabrics with surface areas of 400 cm^2 (53×48 count/cm^2, 140 g/m^2, taffeta fabric) were used as substrates. A non-ionic detergent, $NaHCo_3$, $SnCl_2$, HCl, $PdCl_2$, $AgNO_3$, NH_4OH, NaOH and α-glucose were used in the electroless plating process.

All chemicals were purchased from Merck and used without any further purification.

2.2. Electroless Plating Procedure

Electroless plating was carried out by multi-step processes including scouring, rinsing, sensitization, rinsing, activation, rinsing, electroless silver plating, rinsing and drying. The fabric specimens (20cm × 20 cm) were first scoured in a non-ionic detergent (2g/l) and $NaHCo_3$ (2g/l) solution at 60°C for 30 min prior to use. The samples then were rinsed in distilled water. Next, surface sensitization was conducted by immersion of the samples into an aqueous solution containing $SnCl_2$ (10g/l) and HCl (40ml/l 38% w/w) at 30°C for 20 min. The specimens were again rinsed in distilled water and immersed in an activator containing $PdCl_2$ (0.1g/l) and HCl (20 ml/l, 38% w/w) at 40°C for 5 min. The specimens were rinsed for a third time in a large volume of deionized water for more than 5 min to prevent contamination of the plating bath. Then all samples were immersed in the electroless bath containing silver nitrate (6g/l), ammonium hydroxide as a complexing agent (10ml/l), sodium hydroxide (4g/l) and α-glucose (4g/l) for 30 min at 60°C. The samples were rinsed in hot and cold water respectively, and then were dried in an oven at 70°C.

2.3. Characterization Techniques

The samples were conditioned under standard atmospheric pressure at 65 ± 2% relative humidity and 21 ± 1°C for at least 24 h prior to all measurements.

The changes of fabric weight before and after the plating procedure were determined by a digital weighing scale (HR200, AND Ltd., Japan). The percentage change of fabric weight was calculated by Eq. (1) as follows:

$$\text{Weight change (\%)} = \frac{W - W_0}{W_0} \times 100 \tag{1}$$

The fabric thickness was measured via digital thickness gauge (M034A, SDL Ltd., England) before and after coating. After measurement, the change in thickness can be calculated by Eq. (2) to show the change of fabric thickness.

$$\text{Change in thickness (\%)} = \frac{T - T_0}{T} \times 100 \tag{2}$$

According to ASTM D1388-2001, flexural rigidity of cotton fabric before and after plating was calculated. Stiffness tester (003B, SDL Ltd., England) was employed to measure the length of bending of the fabric in both warp and weft directions. The flexural rigidity was calculated by Eq. (3).

$$G = 0.1 \times M \times C^3 \tag{3}$$

M is the mass of fabric per unit area and C is the bending length.

The tearing strength and elongation of the silver-plated cotton fabric was assessed by (Micro 250, SDL Ltd., England) under constant elongation rate according to ISO 13934-1:1999 standard.

The crease recovery angle of the cotton fabric was measured by crease recovery tester (M003A, SDL Ltd., England) before and after electroless plating according to BS EN 22313:1992 standard.

To study the surface morphology of the cotton fabric before and after plating, the SEM images were recorded by a Philips SEM model 515. The SEM picture of the fabric surface was measured at a magnification of 625 and 10000×. Furthermore, the wave length dispersive x-ray spectrometry (WDX) images were recorded using Microspec, 3PC.

Fourier transform infrared (FTIR) spectra before and after silver coating were acquired by means of Nicolet Magna-IR 560, in the range from 4000 to 500cm^{-1} with 100 scansions and 4 cm^{-1} of band resolution.

The surface resistance of prepared fabric samples was measured by a homemade four-probe apparatus with 1 cm inner distance between four electrodes. Random areas of both the front and back surfaces of the silver-plated fabrics were measured. The measurement carried out was based on ASTM standards F43-93.

The ISO 105-C06:1994 (color fastness to domestic and commercial laundering) standard was used to estimate the effect of washing fastness on the surface resistance and discoloration of plated fabric.

The effect of wet and dry rubbing fastness tests on the color and surface resistance of coated fabric, the electric crockmeter (M238B, SDL Ltd., England) was employed. Wet and dry rubbing fastness was carried out according to ISO 105-X12:2001 standards.

Color fading and surface resistance of coated fabric was assessed by perspiration fastness tests in both acidic and basic solutions under ISO 105-E04:1994 standard conditions.

3. Results and Discussion

3.1. Fabric Characteristics

The fabric weight, thickness and stiffness before and after the coating process were calculated using Formulas 1, 2 and 3, shown in Table 1. The results show that coated fabric was heavier and thicker than the original one. The measured percentage of changes in weight and thickness were 23 and 15, respectively. Thus, it could be demonstrated that silver ions had stuck to the fabric surface impressively. In addition, it could be demonstrated that the silver-plated cotton fabric was stiffer than the untreated one. Bending of the fabric depended on surface friction between the fabric yarns and fibers. After electroless silver plating, the presence of the silver particles on the fibers surface increased the friction between fibers and restricted their relative movement. In consequence, the stiffness of the cotton fabric would increase after silver coating.

Table 1. Measurement of fabric weight, thickness and bending

Specimen (10cm×10cm)	Weight(g)	Thickness(mm)	Flexural rigidity (mg.cm)	
			Warp	Weft
Before plating	1.4	0.4	103.8	68.8
After plating	1.72(↑23%)	0.46(↑15%)	112(↑7.9%)	88.6(↑28.8%)

The estimation of fabric tearing strength, tearing elongation and crease are illustrated in Table 2. According to the obtained results, the tearing strength and tearing elongation of the cotton fabric decreased and increased, respectively, after silver plating. The decreasing of tearing strength and increasing of tearing elongation are the consequence of fabric exposure to highly low pH medium during sensitization and the activation process. The cotton fiber interbonding will be damaged in low pH value resulting in the weakness of the cotton fabric. The values of the crease recovery angle shows a significant improvement for the crease resistance property of cotton fabric after silver electroless plating in both wrap and weft direction. Although the friction between fabric yarns and fibers increased after plating and it seems that the crease recovery angle should decrease, the cotton fabric presence of silver particle on the surface of fibers prevented the creation of the hydrogen bond between cotton fibers during wrinkle implementation. According to obtained results, silver plating on cotton fabric decreases retained crease and enhances wrinkle recovery of the cotton fabric.

Table 2. Estimation of fabric tearing strength, tearing elongation and crease

	Tearing strength(kg.f)	Tearing elongation(mm)	Crease Recovery(degree°)	
			Warp	Weft
Before plating	1667.1	5.47	35	30
After plating	1553.7(↓6.8%)	8.05(↑47.2%)	40(↑14.3%)	32(↑6.67%)

Table 3 summarized the results of estimation of color change after washing, rubbing and perspiration. According to the obtained results, washing and perspiration have no influence on the color of plated fabric. The color fastness results of washing and perspiration are rated as grade 5. The result of rubbing fastness reveals that color of coated fabrics partially changed after dry rubbing. This showed that some silver particles were detached during dry rubbing; likewise, the results show wet rubbing fastness of the plated fabric was comparatively poor considering commercial necessity.

Table 3. Estimation of color change

Washing	Rubbing		Perspiration	
	Dry	Wet	pH 5	pH 8
5	3-4	3	5	5

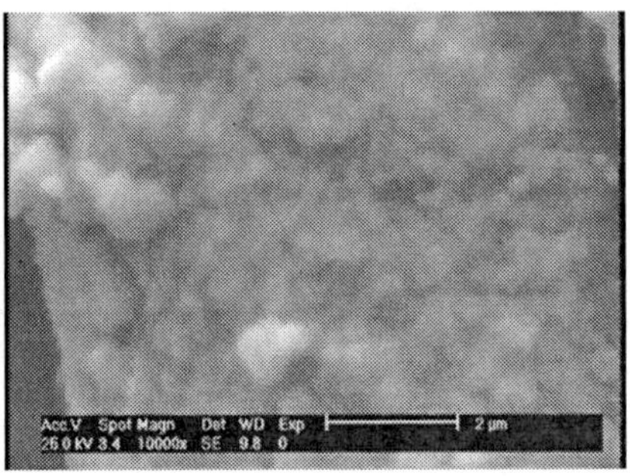

3.2. Fabric Resistance

The surface resistance measurement showed that the resistance of fabric was changed from ∞ to approximately 5.2×10^{-4} Ω/sq which illustrated a remarkable change in surface resistance of silver electroless plated cotton fabric. Results reported in previous studies [1-10], show that surface resistivity of cotton fabric after silver electroless plating was changed from ∞ to 14.9MΩ; in their research, the cotton surface was activated by stannous chloride.

In this work, catalytic sites on the fabric surface were created by proceeding the activation process in which palladium ions are reduced on the surface by stannous ions to a palladium layer that act as a catalyst stead stannous layer for the ensuing electroless deposition of silver. Therefore, in comparison of results, significant development in the surface resistance of electroless silver plating on the cotton fabric could be illustrated by using palladium chloride as an activator agent stead stannous chloride.

The effect of washing on the surface resistance of plated fabric is illustrated in Figure 1. As shown in Figure 1, although the surface resistance of plated fabric after each washing repetition became less, the slope of resistance increasing is downward; furthermore, the change in surface resistance of the coated fabric was negligible.

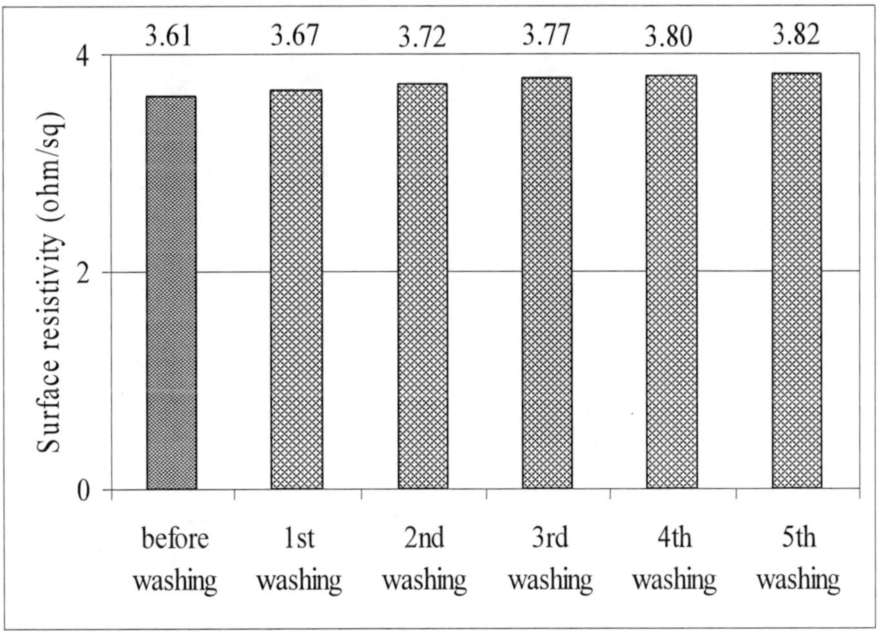

Figure 1. Effect of washing on the surface resistance of silver-plated cotton fabric.

Figure 2 shows the influence of both dry and wet rubbing on the surface resistance of silver-plated cotton fabric. According to the results shown in Figure 2, by reiteration of both dry and wet rubbing tests, the difference of surface resistance between the first and fifth test is insignificant, but meaningful differences appeared in surface resistance of silver-plated fabric observed after first dry and wet rubbing tests compared to the untested one. The surface resistance of coated fabric before the rubbing test was 3.61Ω/sq which decreased after dry and wet rubbing tests to 4.74Ω/sq and 6.10Ω/sq, respectively, which indicates detaching of silver particles from the fabric surface in the first rubbing test; nevertheless, the silver-plated cotton fabric has considerable surface resistance than the uncoated one.

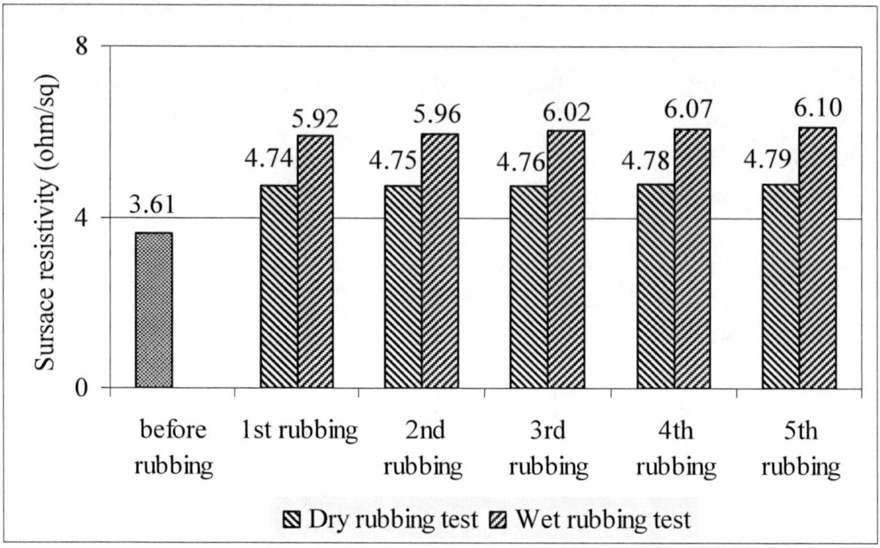

Figure 2. Influence of rubbing on the surface resistance of silver-plasted cotton fabric.

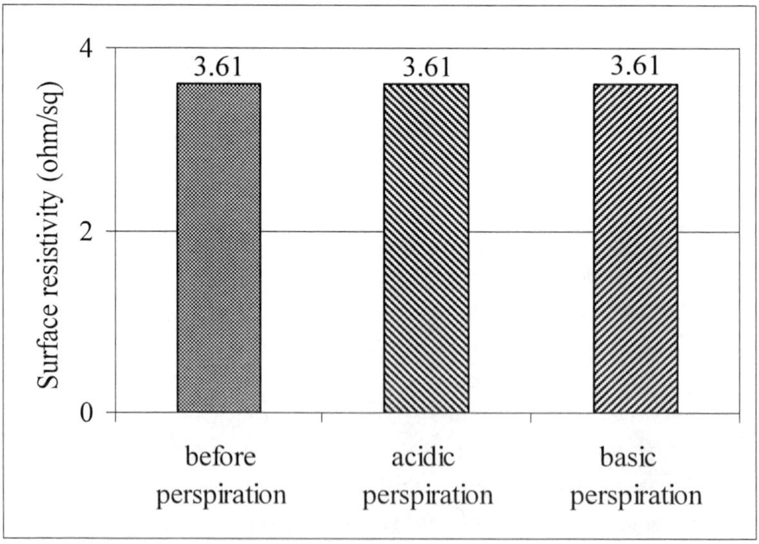

Figure 3. Effect of perspiration on the surface resistance of silver-plated cotton fabric.

In the case of perspiration affect on the surface resistance of plated fabric, according to Figure 3, the obtained results confirm that perspiration influence on surface resistance in both acidic and basic condition was inconsiderable as the surface resistance difference of silver-plated cotton fabric after both acidic and basic perspiration has no change than the untested one.

CONCLUSION

In this study, the cotton fabric was metallized by means of silver electroless plating in which palladium ions act as catalytic sites. Different standard testing methods were employed in order to evaluate the textile characteristics of silver-plated cotton fabric including measurement of weight change, fabric thickness, tearing stress, tearing elongation, flexural rigidity, wrinkle recovery, color fastness to both dry and wet rubbing, color fastness to washing and color fastness to perspiration. Also, the effect of washing, perspiration and both dry and wet rubbing on the surface resistance of silver-deposited cotton fabric was estimated.

The surface resistance of electroless silver-plated cotton fabric was developed significantly in comparison of literature reports by using palladium chloride as the activator agent. In the case of physical properties, the fabric weight and thickness were increased after silver plating. The crease recovery angle and tearing elongation of the metallized fabric were improved. In addition, there were no changes in color fastness to washing and color fastness to perspiration. On the contrary, color fastness to dry and wet rubbing were relatively poor. Silver electroless plating adversely affected the bending and tearing stress of the cotton fabric. The influence of washing, perspiration and both dry and rubbing tests on the surface resistance of silver-deposited cotton fabric were negligible. In consequence of obtained results, it seems that the electroless silver-plated cotton fabric has wonderful application possibility.

ACKNOWLEDGMENT

The authors wish to acknowledge the financial support from The Guilan Science and Technology Park for this work.

References

[1] Ziabari, M., Mottaghitalab, V., McGovern,S.T., Haghi, A.K. *Chim. Phys. Lett.*, 25, 3071 (2008).
[2] Ziabari, M., Mottaghitalab, V., McGovern, S.T., Haghi, A.K. *Nanoscale Research Letter*, 2, 297 (2007).
[3] Ziabari, M., Mottaghitalab, V., Haghi, A.K. *Korean J. Chem. Eng.*, 25, 919 (2008).
[4] Ziabari, M., Mottaghitalab, V., Haghi, A.K. *Korean J. Chem. Eng.*, 25, 923 (2008).
[5] Ziabari, M., Mottaghitalab, V., A.K. *Korean J. Chem. Eng.*, 25, 905 (2008).
[6] Haghi, A.K., Akbari, M. *Physica Status Solidi*, 204, 1830 (2007).
[7] Tan, S.H., Inai, R. *Polymer*, 46, 6128 (2005).
[8] Deitzel, J.M. *Polymer*, 42, 8163 (2001).
[9] Zarkoob, S. *Polymer*, 45, 3973 (2004).
[10] Atheron, S. *Polymer*, 45, 2017 (2004).

In: Development of Nanotechnology in Textiles ISBN: 978-1-62081-030-9
Editors: A. K. Haghi and G. E. Zaikov ©2012 Nova Science Publishers, Inc.

Chapter 2

CARBON/CHITOSAN NANOCOMPOSITES

V. Mottaghitalab and A. K. Haghi[1*]
[1] University of Guilan, Rasht, Iran

INTRODUCTION

At present, terms like "nanobiocomposites" or "biopolymer nanocomposites" are most frequently observed in environmentally friendly research studies. The synthetic polymers have been widely used in various applications of nanocomposites. However, they become a major source of waste after use, because of their poor biodegradability. On the other hand, most of the synthetic polymers show no biocompatibility in *in vivo* and *in vitro* environments. Hence, scientists were interested in biopolymers as biodegradable materials [1], so that several groups of natural biopolymers, such as polysaccharides, proteins, and nucleic acids, came to be used in various applications [2]. Nevertheless, the use of these materials has been limited, because of their relatively poor mechanical properties. Therefore, research efforts have been made to improve the properties of biopolymers as a matrix by means of reinforcement techniques [3].

Chitosan (CHT) is a polysaccharide biopolymer widely used as a matrix in nanobiocomposites, due to its high biocompatibility and biodegradability [4]. Numerous studies have focused on improving the physical properties of

---

* Haghi@Guilan.ac.ir.

biopolymer nanocomposites by using the fundamental behavior of carbon nanotubes [5].

The present study summarizes the recent advances in the production of carbon nanotubes/chitosan nanocomposites by several methods, special stress being laid on the preparation of CNTs/CHT nanofiber composites by the electrospinning method.

1. BIOPOLYMERS

Biomaterials have been defined as materials biocompatible with the living systems. Biocompatibility implies a chemical, physical (surface morphology) and biological suitability of an implant surface to the host tissues. Over the last 30 years, Ramakrishna *et al.* reviewed various biomaterials and their applications, as well as the utilizations of biopolymers and of their biocomposites for medical purposes [6]. These materials can be classified into natural and synthetic biopolymers. Synthetic biopolymers are cheaper and possess high mechanical properties. However, their low biocompatibility, compared with that of natural biopolymers, such as polysaccharides, lipids, and proteins, oriented the attention towards the latter. On the other hand, natural biopolymers usually have weak mechanical properties. Therefore, many efforts have been made to improve them by blending with some filler [7].

Among natural biopolymers, polysaccharides seem to be the most promising materials for various biomedical fields. They come from various sources of animal, plant, algal and microbial origin. Among polysaccharides, chitosan is the most often applied, due to its chemical structure [8].

1.1. Chitosan

Chitin (Fig. 1), the second most abundant natural polymer in the world, is extracted from various plants and animals [9]. Also, derivatives of chitin have been noticed, because of its insolubility in aqueous media. Chitosan (Figure 2) is deacetylated by the derivation of chitin with free amine. Unlike chitin, chitosan is soluble in diluted and organic acids. The polysaccharides contain 2-acetamido-2-deoxy-β-D-glucose and 2-amino-2-deoxy-β-D-glucose. Deacetylation of chitin converts the acetamide groups into amino groups [10]. The deacetylation degree (DD), an important parameter defining chitosan properties, has been defined as "the mole fraction of deacetylated units in the polymer chain" [11]. Chitosan could be suitably modified to impart the desired properties, due to the presence of amino groups. Hence, a wide variety of applications have been reported in recent decades. Table 1 shows chitosan applications in various fields and their main characteristics. The high biocompatibility [12] and biodegradability [13] of chitosan recommend it for potential applications in biomedical fields [14].

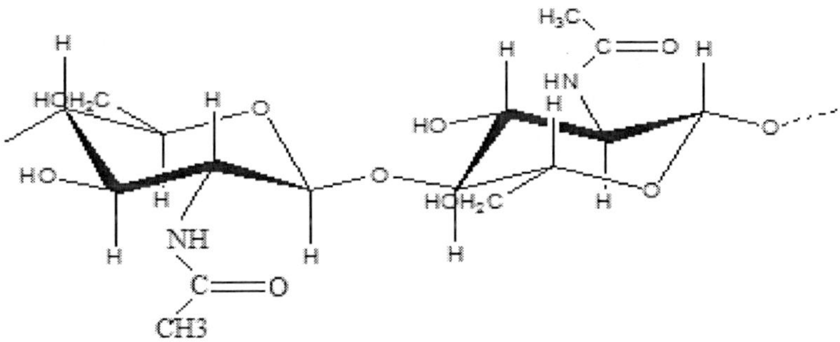

Figure 1. Structure of chitin.

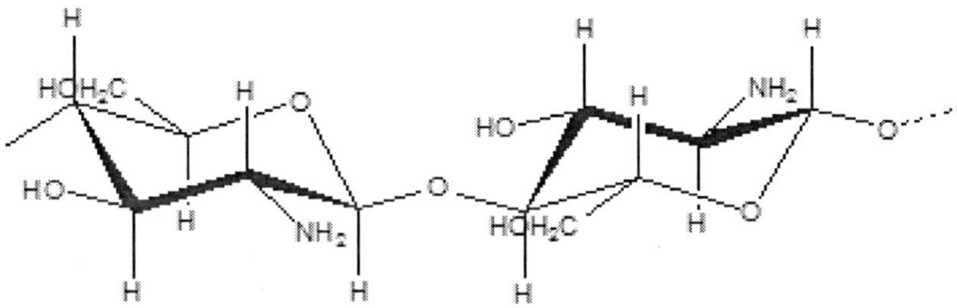

Figure 2. Structure of chitosan.

1.2. Nanobiocomposites with Chitosan Matrix

Chitosan biopolymers have a great potential in biomedical applications, due to their biocompatibility and biodegradability. However, the low physical properties of chitosan limits their applications. The development of high performance chitosan biopolymers involves the incorporation of fillers that display significant mechanical reinforcement [30].

Polymer nanocomposites have been reinforced by nano-sized particles with a high surface area to volume ratio, including nanoparticles, nanoplatelets, nanofibers and carbon nanotubes. Nowadays, carbon nanotubes are considered potential fillers, as they improve the properties of biopolymers [31]. Based on such reports, the researchers assessed the effect of CNT fillers in the chitosan matrix, and evidenced the appropriate properties of CNT/chitosan nanobiocomposites with a high potential for biomedical applications.

Table 1. Chitosan applications in various fields and its main characteristics

Chitosan application		Main characteristics	Ref.
Water engineering		metal ionic adsorption	[15]
Biomedical application	biosensors and immobilization of enzymes and cells	biocompatibility, biodegradability to harmless products, non-toxicity, antibacterial properties, gel-forming properties and hydrophilicity, remarkable affinity to proteins	[16]
Biomedical application	antimicrobial and wound dressing	wound-healing properties	[17]
Biomedical application	tissue engineering	biocompatibility, biodegradable, and antimicrobial properties	[18]
Biomedical application	drug and gene delivery	biodegradable, non-toxicity, biocompatibility, high charge density, mucoadhesion	[19]
Biomedical application	orthopedic/periodontal application	antibacterial properties	[20]
Photography		resistance to abrasion, optical characteristics, film-forming ability	[21]
Cosmetic application		fungicidal and fungi static properties	[22]
Food preservative		biodegradability, biocompatibility, antimicrobial activity, non-toxicity	[23]
Agriculture		biodegradability, non-toxicity, antibacterial properties, cell activator, disease and insect resistant ability	[24]
Textile industry		microorganism resistance, absorption of anionic dyes	[25]
Paper finishing		high density of positive charge, non-toxicity, biodegradability, biocompatibility, antimicrobial and antifungal properties	[26]
Solid-state batteries		ionic conductivity	[27]
Chromatographic separations		presence of free -NH2, primary -OH, secondary -OH	[28]
Chitosan gel for LED and NLO applications		dye-containing chitosan gels	[29]

2. CARBON NANOTUBES

Carbon nanotubes (appendix A), which are tubulars of Buckminster fullerene, were discovered [32] by Iijima in 1991. They are straight segments of tube with arrangements of carbon hexagonal units [33-34]. In recent years, scientists have paid great attention to CNTs, due to their superior electrical, mechanical and thermal properties [35]. Carbon nanotubes are classified as single-walled carbon nanotubes (SWNTs) formed by a single graphene sheet, and multiwalled carbon nanotubes (MWNTs) formed by several graphene sheets wrapped around the tube core [36]. The typical range of diameters of the carbon nanotubes are a few nanometers (~0.8-2 nm at SWNTs [37-38] and ~10-400 nm at MWNTs [39], respectively), and their lengths are up to several micrometers [40]. There are three significant methods for synthesizing CNTs, including arc-discharge [41], laser ablation, [42], chemical vapor deposition (CVD) [43]. The production of CNTs can be also realized by other synthesis techniques, such as the substrate [44], the sol-gel [45] and the gas-phase metal catalyst [46].

The C−C covalent bonding between the carbon atoms is similar to that of the graphite sheets formed by sp^2 hybridization. As a result of this structure, CNTs exhibit a high specific surface area (about 10^3) [47] and, consequently, high tensile strength (more than 200 GPa) and elastic modulus (typically, 1-5 TPa) [48]. The carbon nanotubes also show very high thermal and electrical conductivity. However, these properties differ as to the synthesis methods employed, defects, chirality, degree of graphitization and diameter [49]. For instance, depending on chirality [50], the CNT can be metallic or semiconducting.

The preparation of CNT solutions is impossible, because of their poor solubility. Also, a strong van der Waals interaction of CNTs among several nanotubes leads to aggregation into bundles and ropes [51]. Therefore, various chemical and physical modification strategies will be necessary for improving their chemical affinity [52]. There are two approaches to the surface modification of CNTs, including covalent (grafting) and non-covalent bonding (wrapping) of the polymer molecule onto the surface of CNTs [53]. In addition, the reported cytotoxic effects of CNTs *in vitro* may be mitigated by chemical surface modification [54]. On the other hand, other studies show that the end-caps on nanotubes are more reactive than the sidewalls. Hence, the adsorption of polymers onto the CNT surface can be utilized, together with the functionalization of defects and associated carbons [55].

Chemical modification of CNTs by covalent bonding is one of the important methods for improving their surface characteristics. Because of the extended π-network of the sp^2-hybridized nanotubes, CNTs have a tendency for covalent attachment, which introduces the sp^3-hibrydized C atoms [56]. These functional groups can be attached to the termini of tubes by surface-bound carboxylic acids (grafting), or by direct sidewall modifications of CNTs based on "*in situ* polymerization processing" (grafting form) [57]. Chemical functionalization of CNTs creates various activated groups (such as carboxyl [58], amine [59], fluorine [60], etc.) onto the CNT surface by covalent bonds. However, there are two disadvantages for these methods. First, the CNT structure may be decomposed, due to the functionalization reaction [61] and to a long ultrasonication process [62]. Disruption of the π electron system is reduced as the result of these damages, leading to a decrease of the electrical and mechanical properties of CNTs. Secondly, the acidic and oxidation treatments often used for the functionalization of CNTs are environmentally unfriendly [63]. Thus, non-covalent functionalization of CNTs is largely recommended, as it preserves their intrinsic properties, while improving solubility and processability. According to this method, the non-covalent interaction between the π electrons of the sp^2 hybridized structure at the sidewalls of CNTs and other π electrons is formed by π-π stacking [64]. These non-covalent interactions can appear between CNTs and the amphiphilic molecules (surfactants) (Figure 3a) [65], polymers [66], and biopolymers, such as DNA [67], polysaccharides [68], etc. In the first method, surfactants – including non-ionic surfactants, anionic surfactants and cationic surfactants – are applied for CNT functionalization. The hydrophobic parts of the surfactants are adsorbed onto the nanotube surface and the hydrophilic parts interact with water [69]. Polymers and biopolymers can functionalize CNTs by two methods, including the endohedral one (Figure 3b) and wrapping (Figure 3c). The endohedral method is a strategy for CNT functionalization. According to this method, nanoparticles – such as proteins and DNA – are entrapped in the inner hollow cylinders of CNTs [70]. In another technique, the van der Waals interactions and π-π stacking between CNTs and polymer lead to wrapping of the polymer around the CNTs [71]. Various polymers and biopolymers, such as polyaniline [72], DNA [73] and chitosan [74], interact physically through wrapping of the nanotube surface and π-π stacking by the solubilized polymeric chain. However, Jian *et al.* [75] created a technique for the non-covalent functionalization of SWNTs most similar to π-π stacking by PPE, without polymer wrapping.

Such functionalization methods can provide many applications of CNTs, the most important one being in biomedical science, as biosensors [76], for drug delivery [77] and tissue engineering [78].

2.1. Nanotube Composites

Considering the low physical characteristics of biopolymers, fillers are recommended for the reinforcement of their electrical, mechanical and thermal properties. Following the discovery of CNTs, many efforts have been made for their application as fillers in other polymers, for improving the properties of the matrix polymer [79]. In 1994, Ajayan was the first one to apply CNTs as a filler in epoxy resin, by the alignment method [80]. Later on, numerous studies have focused on CNTs as excellent substitutes for conventional nanofillers in nanocomposites. Recently, numerous polymers and biopolymers have been reinforced by CNTs. As already mentioned, these nanocomposites have remarkable characteristics, compared to the bulk materials, due to their unique properties [81].

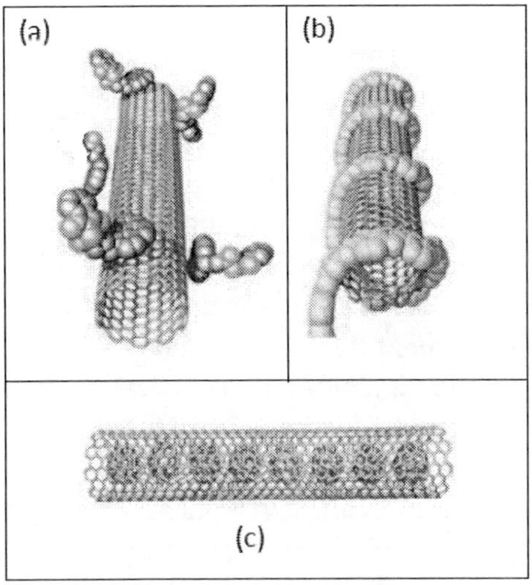

Figure 3. Non-covalent functionalization of CNTs by (a) surfactants, (b) wrapping, (c) the endohedral method.

Several parameters affect the mechanical properties of the composites, including a proper dispersion, a large aspect ratio of the filler, interfacial stress transfer, a good alignment of reinforcement, and solvent selection [82].

The uniformity and stability of nanotube dispersion in polymer matrices are most important parameters for the performance of composites. A good dispersion leads to efficient load transfer concentration centers in composites and to uniform stress distribution [83]. Pemg-Cheng Ma et al. reviewed the dispersion and functionalization techniques of carbon nanotubes for polymer-based nanocomposites, as well as their effects on the properties of CNT/polymer nanocomposites. They demonstrated that the control of these two factors led to uniform dispersion. Overall, the results showed that a proper dispersion enhanced a variety of mechanical properties of nanocomposites [71].

The fiber aspect ratio, defined as "the ratio of average fiber length to fiber diameter", is one of the main effective parameters on the longitudinal elastic modulus [84]. Generally, the carbon nanotubes have a high aspect ratio, but their ultimate performance in a polymer composite is different. The high aspect ratio of the dispersed CNTs could lead to a significant load transfer [85]. However, the aggregation of nanotubes could lead to a decrease in the effective aspect ratio of the CNTs which is one of the most important processing challenges raised by poor CNT dispersion [86].

The interfacial stress transfer has been performed by applying external stresses to the composites. The assessments showed that fillers take a significantly larger share of the load, due to the CNTs-polymer matrix interaction. Also, the literature on the mechanical properties of polymer nanotube composites explains the enhancement of Young's modulus by CNT addition [87]. Wagner et al. investigated the effect of stress-induced fragmentation of multiwalled carbon nanotubes in a polymer matrix. The results showed that polymer deformation generates tensile stress, which is transmitted to CNTs [88].

A homogeneous CNT/polymer matrix alignment in the composite is another effective parameter of carbon nanotube composites. Some researchers [89] assessed the effects of CNT alignment on electrical conductivity and mechanical properties of SWNT/epoxy nanocomposites. The electrical conductivity, Young's modulus and tensile strength of the SWNT/epoxy composites increase with increasing SWNT alignment, due to an increased interface bonding of CNTs in the polymer matrix.

In 2007, Umar Khan et al. examined the effect of solvent selection on the mechanical properties of CNT-polymer composites fabricated from double-walled nanotubes and polyvinyl alcohol composites in different solvents, including water, DMSO and NMP. This work shows that solvent selection can have a dramatic effect on the mechanical properties of CNT-polymer composites [90]. Also, a critical CNT concentration was viewed as an optimum improvement in the mechanical properties of nanotube composites [91].

Other parameters influencing the mechanical properties of nanotube composites are the size, crystallinity, crystalline orientation, purity, entanglement, and straightness. Generally, the ideal CNT properties depend on both matrix and type of application [92].

The various functional groups on the CNT surface permit coupling with the polymer matrix. A strong interface between the coupled CNT/polymer creates an efficient stress transfer. As a previous point, stress transfer is a critical parameter for controlling the mechanical properties of a composite. However, the covalent treatment of CNT reduces the electrical [93] and thermal [94-95] properties of CNTs. Finally, these reductions affect the properties of nanotubes.

The matrix polymer can be wrapped around the CNT surface by non-covalent functionalization, which improves the composite properties through various specific interactions. In their turn, these interactions can improve the properties of nanotube composites [96].

In this context, researchers [97] evaluated the electrical and thermal conductivity in CNT/epoxy composites. Figures 4 and 5 show the electrical and, respectively, thermal conductivity in various filler contents, including carbon black (CB), double-walled carbon nanotube (DWNT) and multi-functionalization. The experimental results showed that electrical and thermal conductivity in nanocomposites is improved by the non-covalent functionalization of CNTs.

2.2. Mechanical and Electrical Properties of Carbon Nanotube/Natural Biopolymer Composites

Table 2 provides mechanical and electrical information on a CNT/natural polymer, compared to a natural polymer. The investigations show the higher mechanical and electrical properties of CNT/natural polymers, compared to those of natural polymers.

2.3. Carbon Nanotube Composite Application

In recent years, special attention has been paid to applying nanotube composites in various fields. Researchers [105] reviewed nanotube composites based on gas sensors, known as playing an important role in industry, environmental monitoring, biomedicine, and so forth. The unique geometry, morphology and material properties of CNTs permitted their application in gas sensors. There are many topical studies for biological and biomedical applications of carbon nanotube composites, based on their biocompatibility [106], such as biosensors [107], tissue engineering [95] and drug delivery [108] in biomedical technology. On the other hand, the light weight, mechanical strength, electrical conductivity and flexibility are significant properties of carbon nanotubes for aerospace applications [109].

Researchers [110] overviewed carbon nanotube composite applications, including electrochemical actuation, strain sensors, power harvesting and bioelectronic sensors, discussing the appropriate elastic and electrical properties for using smart nanoscale materials in the synthesis of intelligent electronic structures. In this context, a polyaniline/SWNT composite fiber was previously developed [111], for evidencing its high strength, robustness, good conductivity and pronounced electroactivity. New battery materials [112] were presented, with mention being made of the enhanced performance of artificial muscles [113] by using these carbon nanotube composites.

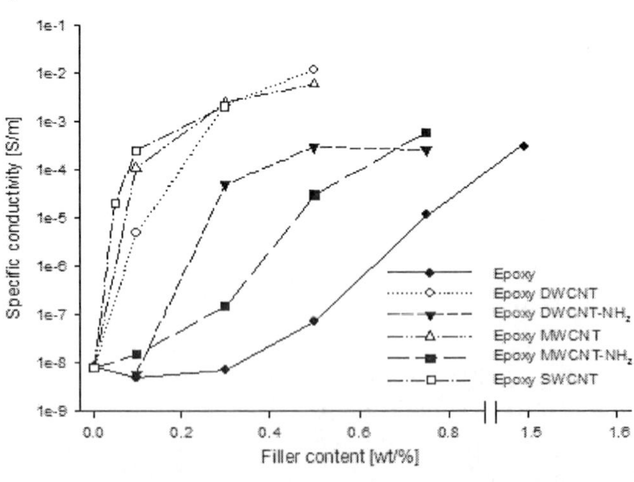

Figure 4. Electrical conductivity of nanocomposites as a function of filler content in weight percent [97].

Table 2. Mechanical and electrical information of neat biopolymers compared to their carbon nanotube nanocomposites

Method	Biopolymer	Mechanical properties					Conductivity (S/cm)	Ref.
		Tensile modulus (Mpa)	Tensile strength (Mpa)	Strain to failure (%)	Comparison modulus (Pa)	Stage modulus (Gpa)		
Polymerized hydrogel	neat collagen				1284±94		$11.37 * 10^{-3}$	[98]
	collagen/CNTs				1127±73		$11.85 * 10^{-3}$	
Solution-evaporation	neat chitosan	1.08±0.04	37.7±4.5				$0.021 * 10^{-9}$	[99]
	chitosan/CNTs	2.15±0.09	74.3±4.6				$120 * 10^{-9}$	
Wet spinning	neat chitosan	4250						[100]
	chitosan/CNTs	1025 0						
Electrospinning	neat silk	140±2.21	6.18±0 3	5.78±0.65			0.028	[101]
	silk/CNTs	4817.24±69.23	44.46±2.1	1.22±0.14			0.144	
Dry-jet wet spinning	neat cellulose	13100±1100	198±25	2.8±0.7		5.1	negligible	[102-103]
	cellulose/CNTs	14900±13 00	257±9	5.8±1.0		7.4	$8.3*10^{-3}$	
Electrospinning	neat cellulose	553±39	21.9±1.8	8.04±0.27				[104]
	cellulose/CNT	1144±37	40.7±2.7	10.46±0.33				

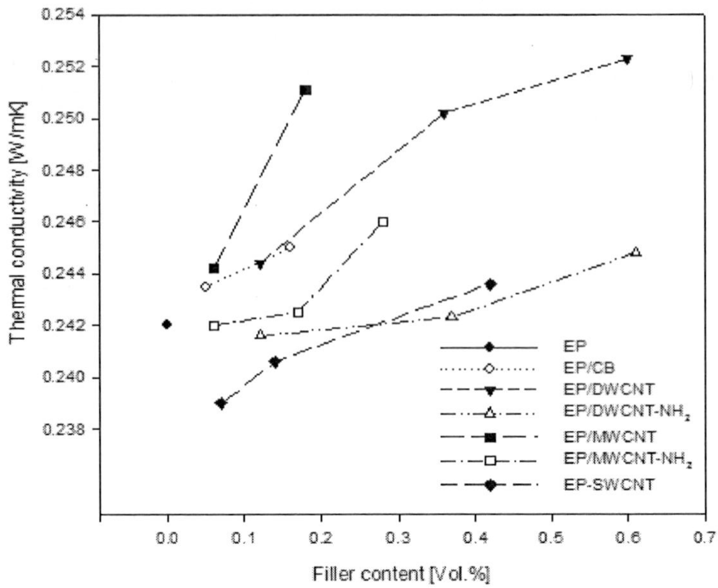

Figure 5. Thermal conductivity as a function of the provided relative interfacial area per gram composite (m^2/g) [97].

Researchers [114] addressed a sustainable environment and green technologies perspective for carbon nanotube applications. These contexts include many engineering fields, such as waste water treatment, air pollution monitoring, biotechnologies, renewable energy technologies and green nanocomposites.

Researchers [115] first discovered the photo-induced electron transfer from CNTs, after which the optical and photo-voltaic properties of carbon nanotube composites have been studied. The results obtained suggested a possible creation of photo-voltaic devices, due to the hole-collecting electrode of CNTs [116].

Food packaging is another remarkable application of carbon nanotube composites. Usually, poor mechanical and barrier properties limit the application of biopolymers. Hence, an appropriate filler is necessary for promoting matrix properties. The unique properties of CNTs have improved thermal stability, strength and modulus, as well as the water vapor transmission rate of the industrially applied composites [117].

3. Chitosan/Carbon Nanotube Composites

The scientists' interest of recent decades in the creation of chitosan/CNT composites is explained by the unique properties of such composites. They attempted to create new properties by adding CNTs to chitosan biopolymers. For example, several articles were published, devoted to their various applications, summarized in the graph plotted in Figure 6.

3.1. Chitosan/Carbon Nanotube Nanofluids

The viscosity and thermal conductivity of nanofluids containing MWNTs stabilized by chitosan were investigated [118]. The MWNT fluid was stabilized with a chitosan solution. The investigations showed that the thermal conductivity enhancements obtained were significantly higher than those predicted by Maxwell's theory. It was also observed that dispersing chitosan into deionized water significantly increased the viscosity of the nanofluid, which explains its non-Newtonian behavior.

3.2. Preparation Methods of Chitosan/CNTs Nanocomposites

There are several methods for the creation of nanobiocomposites, some of them considered for the preparation of chitosan/CNT nanocomposites. These methods are discussed in the following.

3.2.1. Solution-Casting-Evaporation

The electrochemical sensing of a carbon nanotube/chitosan system on dehydrogenase enzymes for preparing glucose biosensors was first assessed [119] in 2004. The nanotube composite was prepared by a solution-casting-evaporation method, according to which the CNT/CHT films were prepared by casting the CNT/CHT solution on the surface of a glassy carbon electrode, and then dried. Such a CNT/CHT system showed a new biocomposite platform for the development of dehydrogenase-based electrochemical biosensors, due to the provided signal transduction of CNT. The highly encouraging results of this composite's utilization in biomedical applications challenged other numerous investigations.

The effect of the CNT/CS matrix on the direct electron transfer of glucose oxidase and glucose biosensor was examined by some researchers [120]. They exhibited high sensitivity and better stability, compared with pure chitosan films. Other investigators [121] used SWNT/CS films for preparing a new galactose biosensor with highly reliable detection of galactose. Some researchers [122] immobilized lactate dehydrogenase within MWNT/CS nanocomposite for producing lactate biosensors, as they provided fast response time and high sensitivity. Also, other investigators [123] demonstrated that immobilization of GOD molecules into a chitosan-wrapped SWNT film is an efficient method for the development of a new class of very sensitive, stable and reproducible electrochemical biosensors.

Several experiments were performed on a DNA biosensor based on chitosan film doped with carbon nanotubes [124]. The observation made was that a CNT/CHT film can be used as a stable and sensitive platform for DNA detection. The results demonstrated improved sensor performance by adding CNT to a chitosan film. Moreover, the analytical performance of glassy carbon electrodes modified with a dispersion of MWNT/CS for DNA quantification was reported [125]. This new platform immobilized the DNA and opened the doors to new strategies for the development of biosensors.

Other experiments [126] reported the high sensitivity of glassy carbon electrodes modified by MWNT-CHT for cathodic stripping voltametric measurements of bromide (Br-).

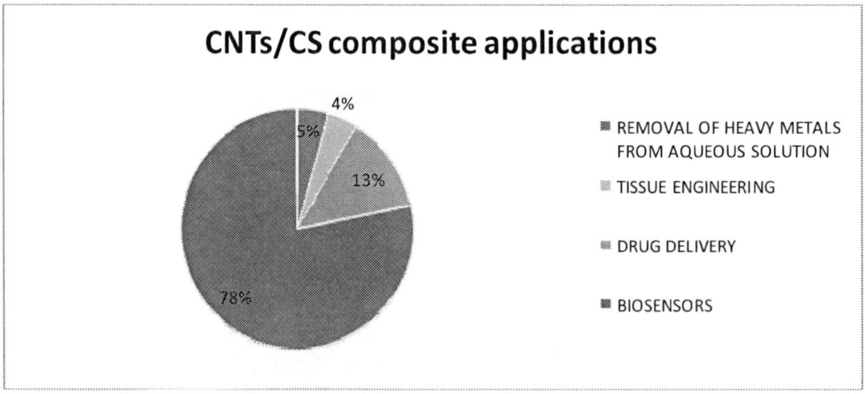

Figure 6. Chitosan/CNT nanocomposite applications.

Qian et al.[127] prepared an amperometric hydrogen peroxide biosensor based on composite films of MWNT/CS. The results showed the excellent electrocatalytical activity of the biosensor for H_2O_2, with good repeatability and stability.

Some researchers [128] reported the effect of the CNT/CHT matrix on an amperometric laccase biosensor, evidencing some of its major advantages, involving detection of different substrates, high affinity and sensitivity, durable long-term stability, and facile preparation procedures.

Others [129] paid particular attention to the preparation of a SWNT/CHT film by the solution-cast method, then characterized their drug delivery properties. They found out that the SWNT/CS film enhanced the slowing down release of dexamethasone.

The growth of apatite on chitosan-multiwalled carbon nanotube composite membranes at low MWNT concentrations was also reported [130]. Apatite was formed on composites with low concentrations.

CNT/CHT nanobiocomposites for immunosensors were produced by some researchers [131]. In such nanobiocomposites, the electron transport enhanced and improved the detection of ochratoxin-A, due to the high electrochemical properties of SWNT. Also, the CNT/CHT nanocomposite used for the detection of human chorionic gonadotrophin antibody was prepared [132], displaying high sensitivity and good reproducibility.

3.2.1.1. Properties and Characterization

The morphology and mechanical properties of chitosan were promoted [99] by adding CNTs. Besides, it was demonstrated [133] that conducting direct electron is very useful for the adsorption of hemoglobin in a CNT/CHT composite film. Studies have demonstrated that this nanobiocomposite can be used in many fields, such as biosensing and biofuel cell applications.

Some researchers [134] evaluated the water transport behavior of chitosan porous membranes containing MWNTs. They characterized two nanotube composites with low molecular weight CSP6K and high molecular weight CSP10K. Because of the hollow nanochannel of MWNTs, located in the pore network of the chitosan membrane, the water transport results for CSP6K enhanced, when the MWNTs content exceeded a critical content. However, for CSP10K series membranes, the water transport rate decreased with increasing the MWNTs content, due to the strong compatibilizing effect of MWNTs.

Other researchers [135] used CNT/CHT nanocomposites with poly(styrene sulfonic acid)-modified CNTs. The thermal, mechanical, and electrical properties of CNT/CHT composite films prepared by solution-casting have potential applications as separation membranes and sensor electrodes.

3.2.2. Crosslinking-Casting-Evaporation

Researchers discovered that MWNTs can be functionalized with –COOH groups at the sidewall defects of nanotubes by carbon nanotubes [136].

According to a novel method [137], chitosan was cross-linked with free –CHO groups by glutaraldehyde and then MWNTs were added to the mixture. The cross-linked MWNT-CHT composite was immobilized with acetylcholinesterase (AChE), for detecting both acetylthiocholine and organophosphorous insecticides. On the other hand, researchers [138] created a new method for cross-linking CHT with carboxylated CNT, involving the addition of glutaraldehyde to the MWNT/CHT solution. They immobilized AChE on the composite for preparing an amperometric acetylthiocholine sensor. The suitable fabrication reproducibility, rapid response, high sensitivity and stability obtained could provide amperometric detection of carbaryl and treazophos [139] pesticide. The results [140] showed the removal of heavy metals, including copper, zinc, cadmium and nickel ions from an aqueous solution in MWNT/CHT nanocomposite films.

3.2.3. Surface Deposition Crosslinking

Researchers [141] decorated carbon nanotubes with chitosan by surface deposition and cros-slinking processes. In this way, the chitosan macromolecules such as polymer cationic surfactants were adsorbed on the CNT surface. In this step, CHT assures a stable dispersion of CNT in an acidic aqueous solution. The pH value of the system was increased by the ammonia solution, to become non-dissolvable of chitosan in aqueous media. Consequently, the soluble chitosan was deposited on the surface of carbon nanotubes similarly to chitosan coating. Finally, the surface-deposited chitosan was cross-linked to CNTs by glutaraldehyde, for potential applications in biosensing, gene and drug delivery of this composite.

3.2.4. The Electro-Deposition Method

Researchers [142] used the nanocomposite film of CNT/CHT as a glucose biosensor by a simple and controllable method. In this one-step electro-deposition method, a pair of gold electrodes was connected to a direct current power supply, and then dipped into the CNT/CHT solution. Herein, the pH near the cathode surface increased, thereby the solubility of chitosan decreased. At a pH of about 6.3, chitosan becomes insoluble and the chitosan-entrapped CNT will be deposited on the cathode surface.

Other investigators [143] characterized electrocatalytic oxidation and sensitive electroanalysis of NADH on a novel film of CS-DA-MWNTs, and improved detection sensitivity. According to this new method, glutaraldehyde cross-linked CHT-DA, with covalent attachment of DA molecules to the CHT chains formed by Schiff bases. Further on, the solution of MWNT dispersed in the CHT-DA solution was dropped on an Au electrode for preparing a CHT-DA-MWNT film, and finally dried.

3.2.5. Covalent Grafting

Carboxylic acid (-COOH) groups were formed on the walls of CNTs by their refluxing in an acidic solution. Carboxylated CNTs were added to the aqueous solution of chitosan. Grafting reactions were accomplished by purging the CNT/CS solution with N_2 and heating to 98 °C. Researchers [144] compared the mechanical properties and water stability of CNTs-grafted-CS with the ungrafted CNTs. A significantly improved dispersion in the chitosan matrix resulted, as well as an important improvement storage modulus and water stability of the chitosan nanocomposites. Researchers [145] created another process for obtaining a CHT-grafted MWNT composite. By this method, after preparing oxidized MWNT (MWNT-COOH), they generated the acyl chloride functionalized MWNT (MWNT-COCl) in a solution of thionyl chloride. In the end, the MWNT-grafted-CS was synthesized by adding CHT to the MWNT-COCl suspension in anhydrous dimethyl formamide. Covalent modification has improved interfacial bonding, resulting in a high stability of the CNT dispersion. Biosensors and other biological applications are evaluated as potential uses of this component. Others [146] prepared a similar composite by reacting CNT-COCl and chitosan with potassium persulfate, lactic acid and acetic acid solution at 75 °C. They estimated that the CNT-grafted-CHT composite can be used in bone tissue engineering, as it may improve the thermal properties.

3.2.5.1. Nucleophilic Substitution Reaction

Covalent modification of MWNT was accomplished with low molecular weight chitosan (LMCS) [147]. This novel derivation of MWNTs can be solved in DMF, DMAc and DMSO, but also in aqueous acetic acid solutions.

3.2.6. Electrostatic Interaction

Some authors [148] synthesized CHT nanoparticles-coated fMWNT composites by electrostatic interactions between the CHT particles and the functionalized CNT. They prepared CHT nanoparticles and CHT microspheres by the precipitation and, respectively, cross-linking method. The electrostatic interactions between CHT particles solved in distilled deionized water and the carboxylated CNTs were confirmed by changing the pH solution. The results obtained showed the same surface charges at pH 2 (both were positively charged) and pH 8 (both were negatively charged). Electrostatic interactions may occur at pH 5.5, due to the different charges between the CHT particles and fCNT, with positive and negative surface charges, respectively. The CHT particles/CNT composite materials could be utilized for potential biomedical applications. Researchers [149] prepared SWNTs/phosphotungstic acid modified SWNT/CS composites using the phosphotungstic acid as an anchor reagent, to modify SWNTs. They succeeded in using PW_{12}-modified SWNT with a negative surface charge, and also positively-charged chitosan, by electrostatic interactions. These strong interfacial interactions between SWNTs and the chitosan matrix showed favorable cyto-compatibility for their potential use as scaffolds for bone tissue engineering.

3.2.7. Microwave Irradiation

Researchers [150] created a new technique for the synthesis of chitosan-modified carbon nanotubes by microwave irradiation. According to this technique, solutions of MWNTs in nitric acid were placed under microwave irradiation and dried for purification of MWNTs. A mixture of purified MWNTs and chitosan solution was reacted in the microwave oven and then centrifuged. The yield black-colored solution was adjusted to pH 8 and centrifuged for precipitation of the CNT/CHT composite. This technique is much more efficient than the conventional methods.

3.2.8. Layer-by-Layer

Researchers [151] characterized MWNT/CHT composite rods with layer-by-layer structure prepared by the *in situ* precipitation method. The samples were prepared by coating the CHT solution on the internal surface of a cylindrical tube and then filling with a MWNT/CHT solution in acetic acid. The morphological, mechanical and thermal properties of this composite rod were also examined.

3.2.8.1. Layer-by-Layer Self-Assembly

Researchers [152] produced a homogeneous multilayer film of MWNT/CHT by the layer-by-layer self-assembly method. According to this method, the negatively charged substrates were dipped into a poly(ethyleneimine) aqueous solution, a MWNT suspension, and a CHT solution, respectively, and finally dried. In this process, both the CHT and PEI solutions contained NaCl for the LBL assembly. The films showed stable optical properties and were appropriate for biosensor applications.

3.2.9. Freeze-Drying

Researchers [153] synthesized and characterized a highly conductive, porous and biocompatible MWNT/CHT biocomposite film by the freeze-drying technique. The process was preformed by freezing a MWNT/CHT dispersion into an aluminum mold, followed by drying.[154]

3.2.10. Wet-Spinning

A recent report [155] showed that chitosan is a good dispersing agent for SWNT. The authors also proposed several methods for preparing a SWNT/CHT macroscopic structure in the form of films, hydrogels and fibers [156]. The CNT/CHT dispersed in acetic acid was spun into an Ethanol:NaOH coagulation solution bath. Better mechanical properties of wet spun fibers resulted from improved dispersion [100].

3.2.11. Electrospinning

Chitosan (CHT)/multiwalled carbon nanotube (MWNT) composite nanofibers can be fabricated by electrospinning [157]. In our experiments, different solvents, including acetic acid 1-90%, formic acid and Tri-Fluoroacetic Acid TFA/DCM, were tested for electrospinning of chitosan/carbon nanotube. No jet was seen when applying a high voltage (even above 25 kV), with 1-30% acetic acid and formic acid as solvent for the chitosan/carbon nanotube.

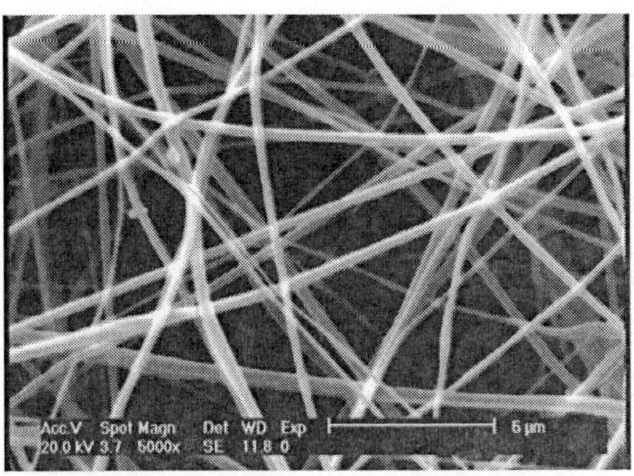

Figure 8. Electron micrograph of electrospun fibers at chitosan concentration of 10 wt%; 24 kV, 5 cm, TFA/DCM: 70/30 [157].

TFA/DCM (70:30) was the only solvent obtained during electrospinning of chitosan/carbon nanotubes. The scanning electron microscopic images showed homogenous fibers with an average diameter of 455 nm (306-672), prepared by dispersing the chitosan/carbon nanotubes in TFA/DCM 70:30. These nanofibers have potential biomedical applications.

Conclusion

In less than ten years, dozens of research studies have been devoted to chitosan biocomposite reinforcements using carbon nanotubes. As a conclusion, much progress has been made in the preparation and characterization of CNT/CHT nanocomposites. Several methods for preparing these nanobiocomposites are reported. In addition, the CNT/CHT applications have been classified, including biomedicine (tissue engineering, biosensors, and drug delivery) and wastewater domains. Numerous potential applications of these nanobiocomposites in tissue engineering and drug delivery are expected. It is believed that an improvement of the preparation methods of CNT/CHT nanocomposites and their characterization will bring a promising future to biomedicine.

Appendix A. Carbon Nanotubes

A.1. Structures of a Crystal Lattice of Diamond and Graphite

In 1991, Japanese researchers studied sediment formed at the cathode during the spray of graphite in an electric arc. Their attention was attracted by the unusual structure of the sediment consisting of microscopic fibers and filaments. Measurements made with an electron microscope showed that the diameter of these filaments do not exceed a few nanometers and a length of from one to several microns.

Having managed to cut a thin tube along the longitudinal axis, the researchers found that it consists of one or more layers, each of which represents a hexagonal grid of graphite, which is based on hexagons with vertices located at the corners of the carbon atoms. In all cases, the distance between the layers is equal to 0.34 nm, which is the same as that between the layers in crystalline graphite.

Typically, the upper ends of tubes are closed by multilayer hemispherical caps; each layer is composed of hexagons and pentagons, reminiscent of the structure of half a fullerene molecule.

The extended structure consisting of rolled hexagonal grids with carbon atoms at the nodes, called nanotubes. Lattice structure of diamond and graphite are shown in Figure A.1 Graphite crystals are built of planes parallel to each other, in which carbon atoms are arranged at the corners of regular hexagons. The distance between adjacent carbon atoms (each side of the hexagon) $d_0 = 0{,}141 nm$, between adjacent planes 0.335 nm.

Each intermediate plane is shifted somewhat toward the neighboring planes, as shown in the figure. The elementary cell of the diamond crystal is a tetrahedron in the center and four vertices of which are carbon atoms. Atoms located at the vertices of a tetrahedron, form a new center of a tetrahedron, and thus, also surrounded by four atoms each, etc. All of the carbon atoms in the crystal lattice are located at equal distance (0.154 nm) from each other. Nanotubes are rolled into a cylinder (hollow tube) graphite plane, which is lined with regular hexagons with carbon atoms at the vertices of a diameter of several nanometers (Figure A.2). Nanotubes can consist of one layer of atoms - single-wall nanotubes SWNT and represent a number of "nested" into one another-layer pipes - multiwalled nanotubes - MWNT. Nanostructures can be collected not only from individual atoms or single molecules, but the molecular blocks. Such blocks or elements to create nanostructures are graphene, carbon nanotubes and fullerenes.

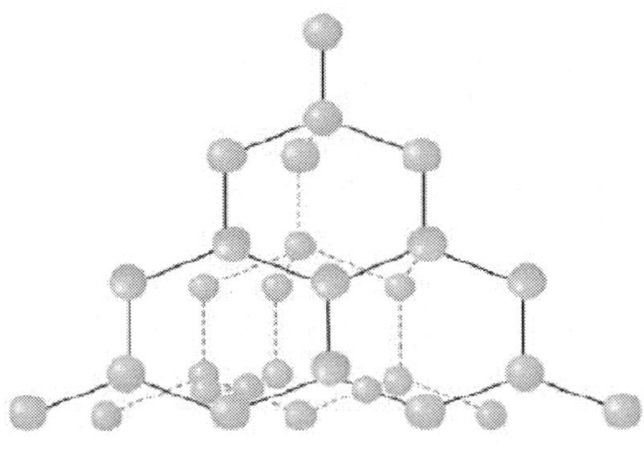

a

Figure A.1. (Continued).

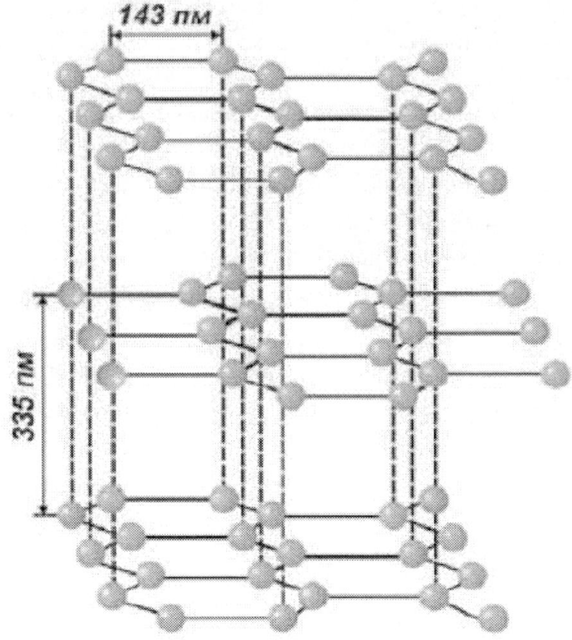

b

Figure A.1. The structure of the diamond lattice a) and graphite b).

A.2. Graphene

Graphene - a single flat sheet, consisting of carbon atoms linked together and forming a grid, each cell is like a bee"s honeycombs (Figure A.2). The distance between adjacent carbon atoms in graphene is about 0.14 nm.

Graphite, from which are made slates of usual pencils, is a pile of graphene sheets (Figure A.3). Graphenes in graphite are very poorly connected and can slide relative to each other. So, if you conduct the graphite on paper, then after separating graphene from sheet the graphite remains on paper. This explains why graphite can write.

A.3. Carbon Nanotubes

Many perspective directions in nanotechnology are associated with carbon nanotubes.

Carbon nanotubes: a carcass structure or a giant molecule consisting only from carbon atoms.

Carbon nanotube is easy to imagine, if we imagine that you fold up one of the molecular layers of graphite - graphene (Figure A.5).

Figure A.2. Schematic illustration of the grapheme.

Figure A.3. Schematic illustration of the three sheets of grapheme.

The way of folding nanotubes - the angle between the directions of nanotube axis relative to the axis of symmetry of graphene (the folding angle) - largely determines its properties.

The way of folding nanotubes - the angle between the directions of nanotube axis relative to the axis of symmetry of graphene (the folding angle) - largely determines its properties.

Figure A.4. Carbon nanotubes.

Of course, no one produces nanotubes, folding it from a graphite sheet. Nanotubes formed themselves, for example, on the surface of carbon electrodes during arc discharge between them. At discharge, the carbon atoms evaporate from the surface and connecting with each other to form nanotubes of all kinds - single, multi-layered and with different angles of twist (Figure A.6).

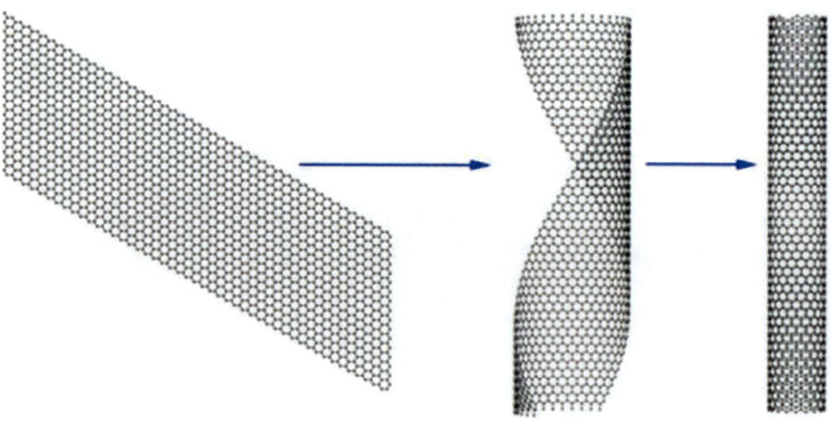

Figure A.5. Imaginary making nanotube (right) from the molecular layer of graphite (left).

The diameter of nanotubes is usually about 1 nm and their length is a thousand times more, amounting to about 40 microns. They grow on the cathode in perpendicular direction to surface of the butt, called self-assembly of carbon nanotubes from carbon atoms. Depending on the angle of folding, the nanotube can have a high conductivity as that of metals and can have properties of semiconductors.

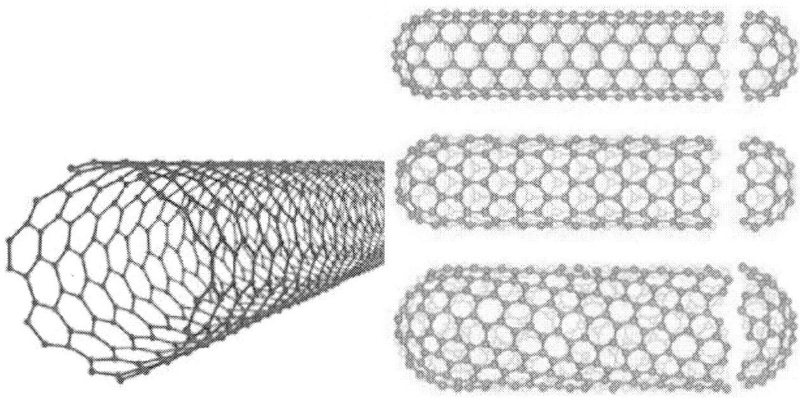

Figure A.6. Left - schematic representation of a single-layer carbon nanotubes, on the right (top to bottom) - two-ply, straight and spiral nanotubes.

Carbon nanotubes are stronger than graphite, although made of the same carbon atoms, because the carbon atoms in graphite are in the sheets. And everyone knows that folding into a tube sheet of paper is much more difficult to bend and break than a regular sheet. That's why carbon nanotubes are strong. Nanotubes can be used as a very strong microscopic rods and filaments, as Young's modulus of single-walled nanotube reaches values of the order of 1-5 TPa, which is much more than steel. Therefore, the thread made of nanotubes, the thickness of a human hair is capable to hold down hundreds of kilos of cargo.

It is true that at present, the maximum length of nanotubes is usually about a hundred microns - which is certainly too small for everyday use. However, the length of the nanotubes obtained in the laboratory, gradually increasing - now scientists have come close to the millimeter border.

A.4. Fullerenes

The carbon atoms, evaporated from a heated graphite surface, connecting with each other, can form not only of the nanotube, but also other molecules, which are closed convex polyhedra, for example, in the form of a sphere or ellipsoid. In these molecules, the carbon atoms located at the vertices of regular hexagons and pentagons that make up the surface of a sphere or ellipsoid.

All of these molecular compounds of carbon atoms, called fullerenes, on behalf of the American engineer, designer and architect R. Buckminster Fuller's domes which were used for construction of its buildings, pentagons and hexagons (Figure A.7), which are the main structural elements of the molecular carcasses of all of fullerenes.

The molecules of the symmetrical and the most studied fullerene consists of 60 carbon atoms (C_{60}), forming a polyhedron consisting of 20 hexagons and 12 pentagons and resembles a soccer ball (Figure A.8). The diameter of the fullerene is about 1 nm.

For the discovery of fullerenes, the American physicist Robert Resins, and the English physicist H. Kroto and R. Curl in 1996 were awarded the Nobel Prize. Many consider the image of the fullerene as a symbol of nanotechnology.

Figure A.7. Biosphere of Fuller (Montreal, Canada).

Figure A.8. Schematic representation of the fullerene C_{60} .

A.5. Classification of Nanotubes

The main classification of nanotubes is conducted by the number of constituent layers.

Single-walled nanotubes: the simplest form of nanotubes. Most of them have a diameter of about 1 nm in length, which can be many thousands of times more. The structure of the nanotubes can be represented as a "wrap" a hexagonal network of graphite (graphene), which is based on hexagons with vertices located at the corners of the carbon atoms in a seamless cylinder. The upper ends of the tubes are closed by hemispherical caps, each layer is composed of six-and pentagons, reminiscent of the structure of half of a fullerene molecule. The distance between adjacent carbon atoms in the nanotube is approximately equal to nm.

Figure A.9. Graphical representation of single-walled nanotube.

Multiwalled nanotubes consist of several layers of graphene stacked in the shape of the tube. The distance between the layers is equal to 0.34 nm, that is the same as that between the layers in crystalline graphite.

Due to its unique properties (high fastness (63 GPa), superconductivity, capillary, optical, magnetic properties, etc.), carbon nanotubes could find applications in numerous areas:

- Additives in polymers;
- Catalysts (autoelectronic emission for cathode ray lighting elements, planar panel of displays, gas discharge tubes in telecom networks);
- Absorption and screening electromagnetic waves;
- Transformation of energy;
- Anodes in lithium batteries;
- Keeping of hydrogen;
- Composites (filler or coating);
- Nanosondes;
- Sensors;
- Strengthening of composites;
- Supercapacitors.

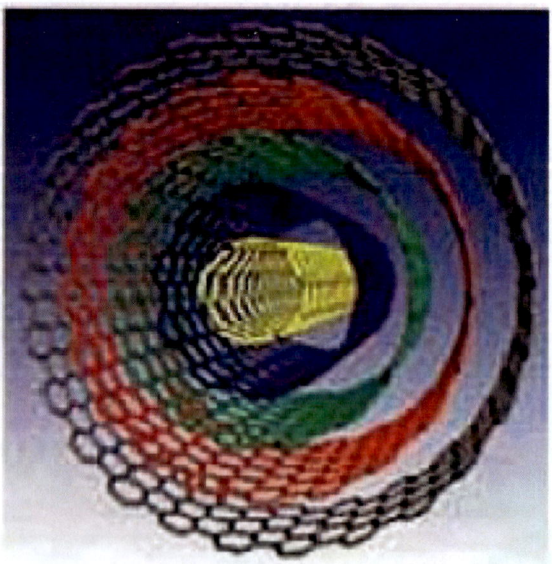

Figure A.10. Graphic representation of a multiwalled nanotube.

For more than a decade, carbon nanotubes, despite their impressive performance characteristics have been used, in most cases, for scientific research. These materials are not yet able to gain a foothold in the market, mainly because of problems with their large-scale production and uncompetitive prices.

To date, the most developed production of nanotubes has been in Asia, the production capacity which is 2-3 times higher than in North America and Europe combined. It is dominated by Japan, which is a leader in the production of MWNT. Manufacturing in North America mainly focused on the SWNT. In the coming years, China will surpass the level of production of the U.S. and Japan, and by now, a major supplier of all types of nanotubes, according to experts, could be South Korea.

A.6. Chirality

Chirality: a set of two integer positive indices (n, m), which determines how the folds the graphite plane and how many elementary cells of graphite at the same time fold to obtain the nanotube.

From the value of parameters, the following are distinguished:

- direct (achiral) high-symmetry carbon nanotubes
  - armchair
  - zigzag or
- helical (chiral) nanotube

In Figure A.11a, a schematic image of the atomic structure of graphite plane – grapheme is shown, and shows how from it, the nanotube can be obtained. The nanotube is folded up with the vector connecting two atoms on a graphite sheet. The cylinder is obtained by folding this sheet so that were combined the beginning and end of the vector. That is, to obtain a carbon nanotube from a graphene sheet, it should turn so that the lattice vector has a circumference of the nanotube in Figure A.11b. This vector can be expressed in terms of the basis vectors of the elementary cell graphene sheet $\vec{R} = n\vec{r_1} + m\vec{r_2}$. Vector $\overline{R}$, which is often referred to simply by a pair of indices (n, m), called the chiral vector. It is assumed that $n > m$. Each pair of numbers represents the possible structure of the nanotube.

In other words, the chirality of the nanotubes indicates the coordinates of the hexagon, which as a result of folding, the plane has to coincide with a hexagon, located at the the beginning of the coordinates (Figure A.12).

Many of the properties of nanotubes (for example, zonal structure or space group of symmetry) strongly depend on the value of the chiral vector. Chirality indicates what property has a nanotube - a semiconductor or metallicheskm. For example, a nanotube (10, 10) in the elementary cell, contains 40 atoms and is the type of metal, whereas the nanotube (10, 9) has1084 atoms and is a semiconductor (Figure A.13).

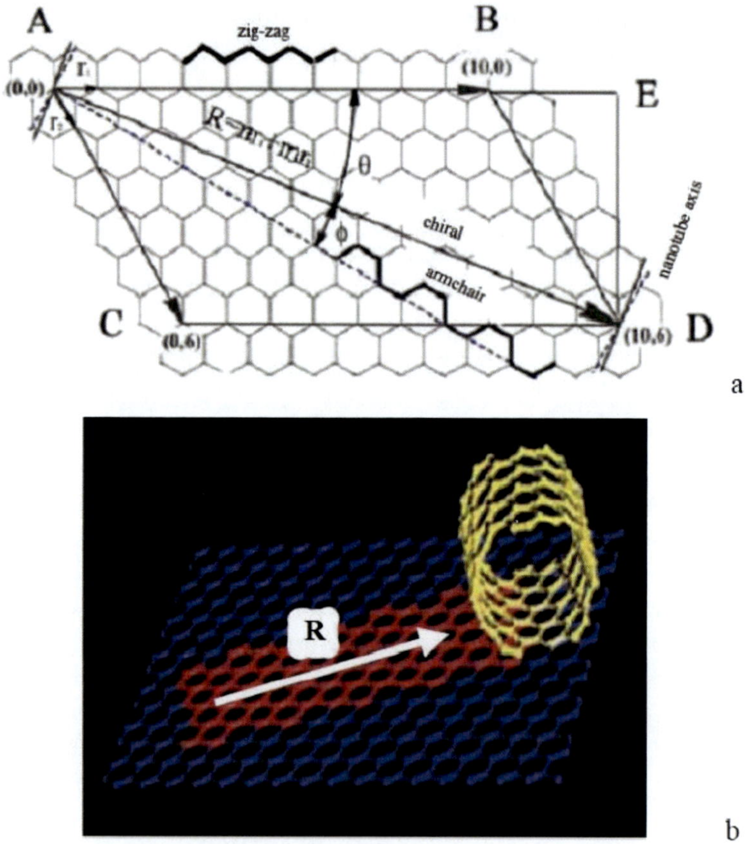

Figure A.11. Atomic structure of graphite plane.

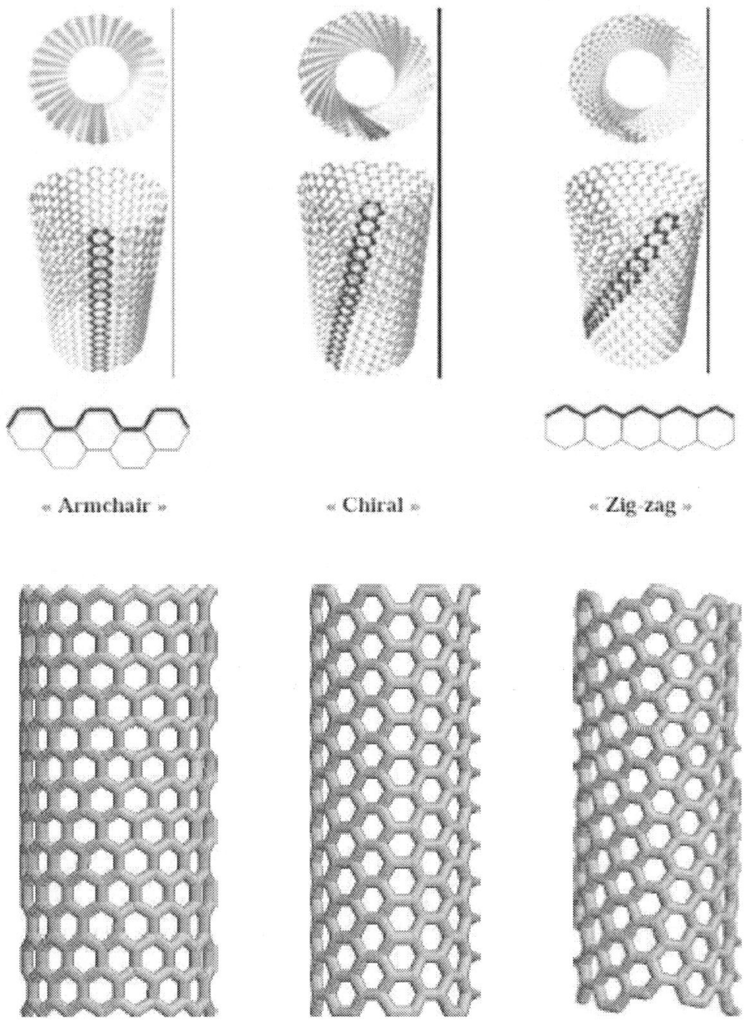

Figure A.12. Single-walled carbon nanotubes in different chirality, Left to right: the zigzag (16, 0), armchair (8, 8) and chiral (10, 6) carbon nanotubes.

If the difference is divisible by 3, then these CNTs have metallic properties. Semimetals are all achiral tubes such as "chair". In other cases, the CNTs show semiconducting properties. Just chair-type CNTs are strictly metal.

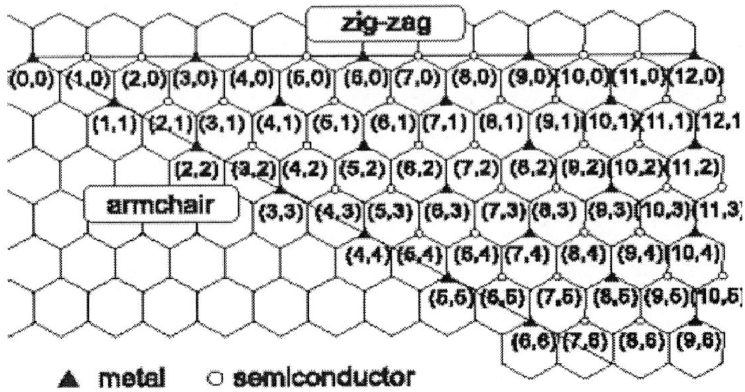

Figure A.13. The scheme of indices of lattice vector tubes having semiconductor and metallic properties.

A.7. Diameter, Chirality Angle and the Mass of Single-Walled Nanotube

Indices of single-walled nanotube chirality unambiguously determine its diameter. Therefore, the nanotubes are typically characterized by a diameter and chirality angle. Chiral angles of nanotubes are the angles between the axis of the tube and the most densely packed rows of atoms. From geometrical considerations, it is easy to deduce relations for the chiral angle and diameter of the nanotube. The angle between the basis vectors of the elementary cell (Figure A.14) and is equal to 60^0 .

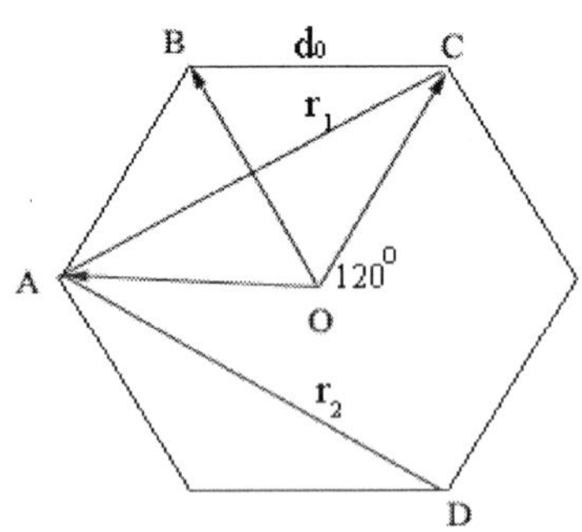

Figure A.14. The elementary cell.

As we know from trigonometry,

$AC^2 = OA^2 + OC^2 - 2OA \cdot OC \cdot \cos 120^0$. As $OA = OC = d_0$, a $r_1 = r_2 = AC$, we have

$$r_1 = r_2 = \sqrt{3} \cdot d_0, \tag{A.1}$$

where - distance between neighboring carbon atoms in the graphite plane. Thus, the basis vectors и of the elementary cell of graphene are $\left|\vec{r_1}\right| = \left|\vec{r_2}\right| = 0{,}244нм$.

Now consider the parallelogram in Figure A.11a.

According to (A.1), we have

$$AB = CD = \sqrt{3}d_0 n,$$

$$AC = BD = \sqrt{3}d_0 m \tag{A.2}$$

Angle $\angle CAB = 60^0$, and $\angle ABD = 120^0$, therefore

$R^2 = 3n^2 d_0^2 + 3m^2 d_0^2 - 2 \cdot 3mnd_0^2 \cos 120^0$, from which we obtain

$R = \sqrt{3}d_0 \sqrt{n^2 + m^2 + mn}$

Taking into account that $R = \pi \cdot d$, then to determine the diameter of the nanotube, we obtain the expression

$$d = \frac{\left|\vec{R}\right|}{\pi} = \sqrt{3\left(m^2 + n^2 + mn\right)} \cdot \frac{d_0}{\pi} \tag{A.3}$$

When we have

$$d = \frac{3nd_0}{\pi}$$

Below in Table A.1, the values of the diameters of nanotubes of different chirality are shown. Thus, knowing the chirality can be found and (possible relations and n, Table A.2). The minimum diameter of the tube is close to 0.4 nm, which corresponds to the chirality (3, 3), (5, 0), (4, 2).

Unfortunately, the objects of that of the diameter of the least stable. Of the single-walled nanotube, it was most stable with chirality indices (10, 10), its diameter is equal 1.35 nm.

We derive a formula for determining the mass of the nanotube with diameter d, length L.

The area of the elementary area is a parallelogram with vertices at the centers of 4 neighboring hexagons (Figure A.15) with the base and height equal to $S_{nл} = \frac{3\sqrt{3}}{2} d_0^2$.

Table A.1. Diameters of nanotubes of different chirality

(n,m)	d, nm	(n,m)	d, nm
(3,2)	0,334	(10,8)	1,232
(4,2)	0,417	(10,9)	1,298
(4,3)	0,480	(11,3)	1,007
(5,0)	0,394	(11,6)	1,177
(5,1)	0,439	(11,10)	1,434
(5,3)	0,552	(12,8)	1,375
(6,1)	0,517	(14,13)	1,844
(7,3)	0,701	(20,19)	2,663
(9,2)	0,801	(21,19)	2,732
(9,8)	1,161	(40,38)	5,326

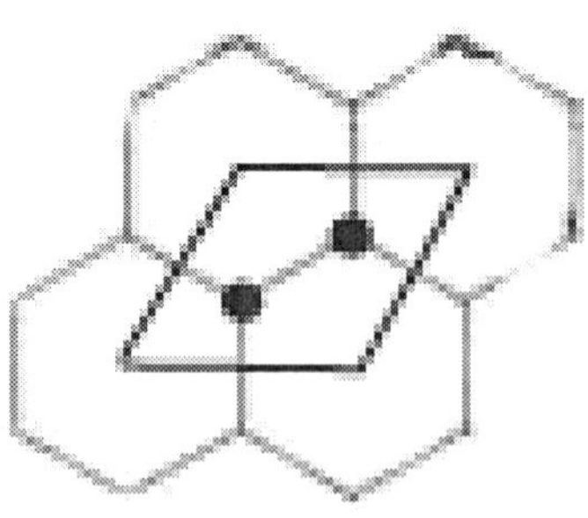

Figure A.15. The elementary area of graphene.

Table A.2. CNT with of different chirality

CNT (n,m)	Diameter CNT, nm	Chirality
(4,0)	0,33	zigzag
(5,0)	0,39	
(6,0)	0,47	
(7,0)	0,55	
(8,0)	0,63	
(9,0)	0,70	
(10,0)	0,78	
(11,0)	0,86	
(12,0)	0,93	
(3,3)	0,40	armchair
(4,4)	0,56	
(5,5)	0,69	
(6,6)	0,81	
(7,7)	0,96	
(8,8)	1,10	
(4,1)	0,39	chiral
(4,2)	0,43	
(7,1)	0,57	
(6,3)	0,62	
(9,1)	0,75	
(10,1)	0,82	
(6,7)	0,90	

The total area of the nanotube is πdL. Consequently, the number of elementary areas is equal $\pi dL / S_{пл}$. At the same time in each elementary site, it contains two carbon atoms. Consequently, the number of carbon atoms in the tube is two times more than the number of elementary areas that can fit on the surface. Therefore, the mass of a carbon nanotube is equal to:

$$m_T = 2m_C \frac{\pi Ld}{S_{пл}} = \frac{4\sqrt{3}\pi \cdot dL}{9d_0^2} m_C, \qquad \text{(A.4)}$$

where $m_C = 12$ - mass of carbon atoms.

To determine the chiral angle from a right triangle, we obtain

$$\sin\theta = \frac{DE}{R},\ \cos\theta = \frac{AE}{R} = \frac{\sqrt{3}nd_0 + BE}{R}$$

If we take into consideration that $\angle EDB = 30^0$, and we see that and $BE = \frac{\sqrt{3}}{2}md_0$ consequently,

$$\sin\theta = \frac{3md_0}{2R},\ \cos\theta = \frac{\sqrt{3}d_0(n + m/2)}{R}$$

From these equalities, we obtain the relation between the chiral indices and angle θ:

$$\theta = arctg\left(\frac{\sqrt{3}m}{2n+m}\right) \tag{A.5}$$

When we have

$$\theta = arctg\frac{\sqrt{3}}{3}$$

References

[1] Praznik, W., Stevens, C.V. in "Renewable Bioresources: Scope and Modification for Non-food Applications", John Wiley & Sons, Ltd, Hoboken, p. 49 (2004).

[2] Parry, D.A.D., Baker, E.N.*Rep. Prog. Phys.*, 47, 1133 (1984).

[3] Bhattacharyya, S., Guillot, S., Dabboue, H., Tranchant, J.F., Salvetat, J.P. *Biomacromolecules,* 9, 505 (2008).

[4] Lavorgna, M., Piscitelli, F., Mangiacapra, P., Buonocore, G.G. *Carbohydr. Polym.*, 82, 291 (2010).

[5] Cao, X., Chen, Y., Chang, P.R., Huneault, M.A. *J. Appl. Polym. Sc.i*, 106, 1431 (2007).

[6] Ramakrishna, S., Mayer, J., Wintermantel, E., Leong, K.W. *Compos. Sci. Technol.*, 61, 1189 (2001).

[7] Liang, D., Hsiao, B.S., Chu, B. *Adv. Drug Delivery Rev.*, 59, 1392 (2007).
[8] Liu, Z., Jiao, Y., Wang, Y., Zhou,C., Zhang, Z. *Adv. Drug Delivery Rev.*, 60, 1650 (2008).
[9] Agboh, O.C., Qin, Y. *Polym. Adv. Technol.*, 8, 355 (1997).
[10] Aranaz, I., Mengíbar, M., Harris, R., Paños, I., Miralles, B., Acosta, N., Galed, G., Heras, A. *Current Chemical Biology*, 3, 203 (2009).
[11] Zhang, Y., Xue, C., Xue, Y., Gao, R., Zhang, X. *Carbohydr. Res.*, 340, 1914 (2005).
[12] VandeVord, P.J., Matthew, H.W.T., DeSilva, S.P., Mayton, L., Wu, B., Wooley, P.H. *J. Biomed. Mater. Res.*, 59, 585 (2002).
[13] Ratajska, M., Strobin, G., Wiśniewska-Wrona, M., Ciechańska, D., Struszczyk, H., Boryniec, S., Biniaś, D., Biniaś, W. *Fibers Text. East Eur.*, 11, 75 (2003).
[14] Jayakumar, R., Prabaharan, M., Nair, S.V., Tamura, H. *Biotechnol. Adv.*, 28, 142 (2010).
[15] Bamgbose, J.T., Adewuyi, S., Bamgbose, O., Adetoye, A.A. *Afr. J. Biotechnol.*, 9, 2560 (2010).
[16] Krajewska, B. *Enzyme Microb. Technol.*, 35, 126 (2004).
[17] Ueno, H., Mori, T., Fujinaga, T. *Adv. Drug Delivery Rev.*, 52, 105 (2001).
[18] Kim, I.-Y., Seo, S.-J., Moon, H.-S., Yoo, M.-K., Park, I.-Y., Kim, B.-C., Cho, C.-S. *Biotechnol. Adv.*, 26, 1 (2008).
[19] Sinha, V.R., Singla, A.K., Wadhawan, S., Kaushik, R., Kumria, R., Bansal, K., Dhawan, S. *Int. J. Pharm.*, 274, 1 (2004).
[20] Martino, A.D., Sittinger, M., Risbud, M.V. *Biomaterials*, 26, 5983 (2005).
[21] Muzzarelli, R.A.A. *Cell Mol. Life Sci.*, 53, 131 (1997).
[22] Muzzarelli, R.A.A., Muzzarelli, C., Tarsi, R., Miliani, M., Gabbanelli, F., Cartolari, M. *Biomacromolecules*, 2, 165 (2001).
[23] Dutta, P.K., Tripathi, S., Mehrotra, G.K., Dutta, J. *Food Chem.*, 114, 1173 (2009).
[24] Boonlertnirun, S., Boonraung, C., Suvanasara, R. *J. Metals Materials Minerals,* 18, 47 (2008).
[25] Huang, K.-S., Wu, W.-J., Chen, J.-B., Lian, H.-S. *Carbohydr. Polym.*, 73, 254 (2008).
[26] Lertsutthiwong, P., Chandrkrachang, S., Stevens, W.F. *J. Metals Materials Minerals*, 10, 43 (2000).
[27] Subban, R.H.Y., Arof, A.K. *Phys. Scr.*, 53, 382 (1996).

[28] Ottoy, M.H., Vårum, K.M., Christensen, B.E., Anthonsen, M.W., Smidsrød, O. *Carbohydr. Polym.*, 31, 253 (1996).
[29] Dutta, P.K., Dutta, J., Tripathi, V.S. *J. Sci. Ind. Res.*, 63, 20 (2004).
[30] Li, Q., Zhou, J., Zhang, L. *J. Polym. Sci. Pol. Phys.*, 47, 1069 (2009).
[31] Thostenson, E.T., Li, C., Chou, T.-W. *Compos. Sci. Technol.*, 65, 491 (2005).
[32] Iijima, S. *Nature*, 354, 56 (1991).
[33] Benthune, D.S., Kiang, C.H., Vries, M.S.D., Gorman, G., Savoy, R., Vazquez, J. Beyers, R. *Nature*, 363, 605 (1993).
[34] Iijima, S., Ichihashi, T. *Nature*, 363, 603 (1993).
[35] Trojanowicz, M. *TrAC Trends Anal. Chem.*, 25, 480 (2006).
[36] Duclaux, L. *Carbon*, 40, 1751 (2002).
[37] Wang, Y.Y., Gupta, S., Garguilo, J.M., Liu, Z.J., Qin, L.C., Nemanich, R.J., *Diamond and Related Materials*, 14, 714 (2005).
[38] Guo, J., Datta, S., Lundstrom, M. *IEEE Transaction On Electron Devices*, 51, 172 (2004).
[39] Kuo, C.-S., Bai, A., Huang, C.-M., Li, Y.-Y., Hu, C.-C., Chen, C.-C. *Carbon*, 43, 2760 (2005).
[40] Jacobsen, R.L., Tritt, T.M., Guth, J.R., Ehrlich, A.C., Gillespie, D.J. *Carbon*, 33, 1217 (1995).
[41] Journet, C., Maser, W.K., Bernier, P., Loiseau, A., Chapelle, M.L.I., Lefrant, S., Deniard, P., Leek, R., Fischer, J.E. *Nature*, 388, 756 (1997).
[42] Thess, A., Lee, R., Nikolaev, P., Dai, H., Petit, P., Robert, J., Xu, C., Lee, Y.H., Kim, S.G., Rinzler, A.G., Colbert, D.T., Scuseria, G.E., Tomanek, D., Fischer, J.E., Smalley, R.E. *Science*, 273, 483 (1996).
[43] Cassell, A.M., Raymakers, J.A., Kong, J., Dai, H. *J. Phys. Chem., B*, 103, 6482 (1999).
[44] Fan, S., Liang, W., Dang, H., Franklin, N., Tombler, T., Chapline, M. Dai, H. *Physica E.*, 8, 179 (2000).
[45] Xie, S., Li, W., Pan, Z., Chang, B., Sun, L. *Mater. Sci. Eng., A*, 286, 11 (2000).
[46] Tang, Z.K., Zhang, L., Wang, N., Zhang, X.X., Wen, G.H., Li, G.D., Wang, J.N., Chan, C.T., Sheng, P. *Science*, 292, 2462 (2001).
[47] Peigney, A., Laurent, C., Flahaut, E., Bacsa, R.R., Rousset, A. *Carbon*, 39, 507 (2001).
[48] Pan, Z.W., Xie, S.S., Lu, L., Chang, B.H., Sun, L.F., Zhou, W.Y., Wang,G., Zhang, D.L. *Appl. Phys. Lett.*, 74, 3152 (1999).
[49] Forro, L., Salvetat, J.P., Bonard, J.M., Basca, R., Thomson, N.H., Garaj, S., Thien-Nga, L., Gaal, R., Kulik, A., Ruzicka, B., Degiorgi, L.,

Bachtold, A., Schonenberger, C., Pekker, S., Hernadi, K. *Science Application Nanotubes*, 297 (2000).
[50] Frank, S., Poncharal, P., Wang, Z.L., Heer, W.A.D. *Science*, 280, 1744 (1998).
[51] Britzab, D.A., Khlobystov, A.N. *Chem. Soc. Rev.*, 35, 637 (2006).
[52] Andrews, R., Weisenberger, M.C. *Current Opin. Solid State Mater. Sci.*, 8, 31 (2004).
[53] Hirsch, A. *Angew. Chem. Int. Ed.*, 41, 1853 (2002).
[54] Firme, C.P., Prabhakar, R.B. *Nanomed. Nanotechnol. Biol. Med.*, 6, 245 (2010).
[55] Niyogi, S., Hamon, M.A., Hu, H., Zhao, B., Bhowmik, P., Sen, R., Itkis, M.E., Haddon, R.C. *Acc. Chem. Res.*, 35, 1105 (2002).
[56] Kuzmany, H., Kukovecz, A., Simona, F., Holzweber, M., Kramberger, C., Pichler, T. *Synth. Met.*, 141, 113 (2004).
[57] Spitalsky, Z., Tasis, D., Papagelis,K., Galiotis, C. *Prog. Polym. Sci.*, 35, 357 (2010).
[58] Narain, R., Housni, A., Lane, L. *J. Polym. Sci., A, Polym. Chem.*, 44, 6558 (2006).
[59] Wang, M., Pramoda, K.P., Goh, S.H. *Carbon*, 44, 613 (2006).
[60] Touhara, H., Yonemoto, A., Yamamoto, K., Komiyama, S., Kawasaki, S., Okino, F., Yanagisawa, T., Endo, M. *Fluorine Chem.*, 114, 181 (2002).
[61] Zhang, W., Sprafke, J.K., Ma, M., Tsui, E.Y., Sydlik, S.A., Rutledge, G., Swager, T.M. *J. Am. Chem. Soc.*, 131, 8446 (2009).
[62] He, P., Gao, Y., Lian, J., Wang, L., Qian, D., Zhao, J., Wang, W., Schulz, M.J., Zhou, X.P., and. Shi, D. *Compos., Part A-Appl. S.*, 37, 1270 (2006).
[63] Sulong, A.B., Azhari, C.H., Zulkifli, R., Othman, M.R., Park, J. *Eur. J. Sci. Res.*, 33, 295 (2009).
[64] Wang, C., Guo, Z.X., Fu, S., Wu, W., Zhu, D. *Prog. Polym. Sci.*, 29, 1079 (2004).
[65] Rausch, J., Zhuang, R.-C., Mäder, E. *Compos., Part A, Appl. S.*, 41, 1038 (2010).
[66] Sahoo, N.G., Rana, S., Cho, J.W., Li, L., Chan, S.H., *Prog. Polym. Sci.*, 35, 837 (2010).
[67] Zheng, D., Li, X., Ye, J. *Bioelectrochemistry*, 74, 240 (2009).
[68] Zhang, X., Meng, L., Lu, Q. *ACS Nano.*, 10, 3200 (2009).
[69] Wang, H. Current Opin. Colloid Int., 14, 364 (2009).

[70] Schlüter, A.D., Hirsch, A Vostrowsky, O. *Functional Molecular Nanostructures*, 245, 193 (2005).
[71] Ma, P.-C., Siddiqui, N.A., Marom, G., Kim, J.-K. *Composites, Part A* 41, 1345 (2010).
[72] Mottaghitalab, V., Spinks, G.M., Wallace, G.G. *Synth. Met.*, 152, 77 (2005).
[73] Cheung, W., Pontoriero, F., Taratula, O., Chen, A.M., He, H. *Adv. Drug Delivery Rev.*, 62, 633 (2010).
[74] Piovesan, S., Cox, P.A., Smith, J.R., Fatouros, D.G., Roldo, M. *Phys. Chem., Chem. Phys.*, 12, 15636 (2010).
[75] Chen, J., Liu, H., Weimer, W.A., Halls, M.D., Waldeck, D.H., Walker, G.C. *J. Am. Chem. Soc.*, 124, 9034 (2002).
[76] Vamvakaki, V.,Fouskaki, M., Chaniotakis, N. *Anal. Lett.*, 40, 2271 (2007).
[77] Kang, Y., Liu, Y.-C., Wang, Q., Shen, J.-W., Wu, T., Guan, W.-J. *Biomaterials*, 30, 2807 (2009).
[78] Harrison, B.S., Atala, A. *Biomaterials*, 28, 344 (2007).
[79] Moniruzzaman, M., Winey, K.I. *Macromolecules*, 39, 5194 (2006).
[80] Ajayan, P.M.,Stephan, O., Colliex, C., Trauth, D. *Science*, 265, 1212 (1994).
[81] Liu, P. *Eur. Polym. J.*, 41, 2693 (2005).
[82] Manchado, M.A.L., Valentini, L., Biagiotti, J., Kenny, J.M. *Carbon*, 43, 1499 (2005).
[83] Wang, Q., Varadan, V.K. *Smart Mater. Struct.,* 14, 281 (2005).
[84] Mallick, P.K.; in "Fiber Reinforced Composites: Materials, Manufacturing, and Design", Taylor & Francis Group, London, p. 43 (2008).
[85] Shaffer, M.S.P., Sandler, J.K.W.; in "Processing and properties of nanocomposites", World Scientific Publishing Co. Pte. Ltd., Singapore, p. 31 (2006).
[86] Nan, C.W., Shi, Z., Lin, Y. *Chem. Phys. Lett.*, 375, 666 (2003).
[87] Coleman, J.N., Khan, U., Blau, W.J., Gun'ko, Y.K. *Carbon*, 44, 1624 (2006).
[88] Wagner, H.D., Lourie, O., Feldman, Y.,Tenne, R. *Appl. Phys. Lett.,* 72, 188 (1998).
[89] Wang, Q., Dai, J., Li, W., Wei, Z., Jiang, J. *Compos. Sci. Technol.*, 68, 1644 (2008).
[90] Khan, U., Ryan, K., Blau, W.J., Coleman, J.N. *Compos. Sci. Technol.,* 67, 3158 (2007).

[91] Allaoui, A., Bai, S., Cheng, H.M., Bai, J.B. *Compos. Sci. Technol.*, 62, 1993 (2002).

[92] Esawi, A.M.K., Farag, M.M. *Mater Design*, 28, 2394 (2007).

[93] Gojny, F.H., Wichmann, M.H.G., Fiedler, B., Kinloch, I.A., Bauhofer, W., Windle, A.H., Schulte, K. *Polymer*, 47, 2036 (2006).

[94] Kamaras, K., Itkis, M.E., Hu, H., Zhao, B., Haddon, R.C. *Science*, 301, 1501 (2003).

[95] Shenogin, S., Bodapati, A., Xue, L., Ozisik, R., Keblinski, P. *Appl. Phys. Lett.*, 85, 2229 (2004).

[96] MacDonald, R.A., Laurenzi, B.F., Viswanathan, G., Ajayan, P.M., Stegemann, J.P. *J. Biomed. Mater. Res., A*, 74A, 489 (2005).

[97] Bose, S., Khare, R.A., Moldenaers, P. *Polymer*, 51, 975 (2010).

[98] Tosun, Z., McFetridge, P.S. *J. Neural Eng.*, 7, 1 (2010).

[99] Wang, S.-F., L. Shen, L., Zhang, W.-D.,Tong, Y.-T. *Biomacromolecules,* 6, 3067 (2005).

[100] Spinks, G.M., Shin, S.R., Wallace, G.G., Whitten, P.G., Kim, S.I., Kim, S.J. *Sens. Actuators B*, 115, 678 (2006).

[101] Gandhi, M., Yang, H., Shor, L., Ko, F. *Polymer*, 50, 1918 (2009).

[102] Rahatekar, S.S., Rasheed, A., Jain, R., Zammarano, M., Koziol, K.K., Windle, A.H., Gilman, J.W., Kumar, S. *Polymer*, 50, 4577 (2009).

[103] Zhang, H., Wang, Z., Zhang, Z., Wu, J., Zhang, J., He, J. *Adv. Mater.*, 19, 698 (2007).

[104] Lu, P., Hsieh, Y.-L. *ACS Appl. Mater. Interfaces,* 2, 2413 (2010).

[105] Wang, Y., Yeow, J.T.W. *J. Sensors,* 2009, 1 (2009).

[106] Yang, W., Thordarson, P., Gooding, J.J., Ringer, S.P., Braet, F. *Nanotechnology,* 18, 1 (2007).

[107] Wang, J. *Electroanal.*, 17, 7 (2005).

[108] Foldvari, M., Bagonluri, M. *Nanomed. Nanotechnol. Biol. Med.*, 4, 183 (2008).

[109] Belluccia, S., Balasubramanianab, C., Micciullaac, F., Rinaldid, G. *J. Exp. Nanosci.*, 2, 193 (2007).

[110] Kang, I., Heung, Y.Y., Kim, J.H., Lee, J.W., Gollapudi, R., Subramaniam, S., Narasimhadevara, S., Hurd, D., Kirikera, G.R., Shanov, V., Schulz, M.J., Shi, D., Boerio, J., Mall, S., Ruggles-Wren, M. *Composites, Part B,* 37, 382 (2006).

[111] Mottaghitalab, V., Spinks, G.M., Wallace, G.G. *Polymer*, 47, 4996 (2006).

[112] Wang, C.Y., Mottaghitalab, V., Too, C.O., Spinks, G.M., Wallace, G.G. *J. Power Sources*, 163, 1105 (2007).

[113] Mottaghitalab, V., Xi, B., Spinks, G.M., Wallace, G.G. *Synth. Met.*, 156, 796 (2006).
[114] Ong, Y.T., Ahmad, A.L., Zein, S.H.S., Tan, S.H. *Braz. J. Chem. Eng.*, 27, 227 (2010).
[115] Sariciftci, N.S., Smilowitz, L., Heeger, A.J., Wudi, F. *Science*, 258, 1474 (1992).
[116] Harris, P.J.F. *Int. Mater. Rev.*, 49, 31 (2004).
[117] Azeredo, H.M.C.D. *Food Res. Int.*, 42, 1240 (2009).
[118] Phuoc, T.X., Massoudi, M., Chen, R.-H. *Int. J. Therm. Sci.*, 50, 12 (2011).
[119] Zhang, M., Smith, A., Gorski, W. *Anal. Chem.*, 76, 5045 (2004).
[120] Liu, Y., Wang, M., Zhao, F., Xu, Z. Dong, S. *Biosens. Bioelectron.*, 21, 984 (2005).
[121] Tkac, J., Whittaker, J.W., Ruzgas, T. *Biosens. Bioelectron.*, 22, 1820 (2007).
[122] Tsai, Y.-C., Chen, S.-Y., Liaw, H.-W. *Sens. Actuators B*, 125, 474 (2007).
[123] Zhou, Y., Yang, H., Chen, H.-Y. *Talanta*, 76, 419 (2008).
[124] Li, J., Liu, Q., Liu, Y., Liu, S., Yao, S. *Anal. Biochem.*, 346, 107 (2005).
[125] Bollo, S., Ferreyr, N.F., Rivasb, G.A. *Electroanal.*, 19, 833 (2007).
[126] Zeng, Y., Zhu, Z.H., Wang, R.X., Lu, G.-H. *Electrochim. Acta*, 51, 649 (2005).
[127] Qian, L., Yang, X. *Talanta*, 68, 721 (2006).
[128] Liu, Y., Qu, X., Guo, H., Chen, H., Liu, B., Dong, S. *Biosens. Bioelectron.*, 21, 2195 (2006).
[129] Naficy, S., Razal, J.M., Spinks,G.M., Wallace, G.G. *Sens. Actuators A*, 155, 120 (2009).
[130] Yang, J., Yao, Z., Tang, C., Darvell, B.W., Zhang, H., Pan, L., Liu, J., Chen, Z. *Appl. Surf. Sci.*, 255, 8551 (2009).
[131] Kaushik, A., Solanki, P.R., Pandey, M.K. Kaneto, K., Ahmad, S., Malhotra, B.D. *Thin Solid Films*, 519, 1160 (2010).
[132] Yang, H., Yuan, R., Chai,Y.,Ying, Z. *Colloids Surf., B*, 82, 463 (2011).
[133] Zheng, W., Chen, Y.Q., Zheng, Y.F. *Appl. Surf. Sci.*, 255, 571 (2008).
[134] Tang, C., Zhang, Q., Wang, K., Fu, Q., Zhang, C. *J. Membr. Sci.*, 337, 240 (2009).
[135] Liu, Y.-L., Chen, W.-H., Chang, Y.-H. *Carbohydr. Polym.*, 76, 232 (2009).
[136] Ghica, M.E., Pauliukaite, R., Fatibello-Filho, O., Brett, C.M.A. *Sens Actuators B*, 142, 308 (2009).

[137] Kandimalla, V.B., Ju, H. *Chem. Eur. J.*, 12, 1074 (2006).
[138] Du, D., Huang, X., Cai, J., Zhang, A.,Ding, J., Chen, S. *Anal. Bioanal. Chem.,* 387, 1059 (2007).
[139] Du, D., Huang, X., Cai, J., Zhang, A. *Sens. Actuators B*, 127, 531 (2007).
[140] Salam, M.A., Makki, M.S.I., Abdelaal, M.Y.A. *J. Alloys Compd.*, 509, 2582 (2010).
[141] Liu, Y., Tang, J., Chen, X., Xin, J.H. *Carbon*, 43, 3178 (2005).
[142] Luo, X._L., Xu, J.-J., Wang, J.-L., Chen, H.-Y. *Chem. Commun.*, 16, 2169 (2005).
[143] Ge, B., Tan, Y., Xie, Q., Ma, M., Yao, S. *Sens. Actuators B*, 137, 547 (2009).
[144] Shieh, Y.-T., Yang, Y.-F. *Eur. Polym. J.*, 42, 3162 (2006).
[145] Wu, Z., Feng, W., Feng, Y., Liu, Q., Xu, X., Sekino, T., Fujii, A., Ozaki, M. *Carbon*, 45, 1212 (2007).
[146] Carson, L., Kelly-Brown, C., Stewart, M., Oki, A., Regisford, G., Luo, Z., Bakhmutov, V.I. *Mater. Lett.*, 63, 617 (2009).
[147] Ke, G., Guan, W.C., Tang, C.Y., Hu, Z., Guan, W.J., Zeng, D.L., Deng, F. *Chin. Chem. Lett.*, 18, 361 (2007).
[148] Baek, S.-H., Kim, B., Suh, K.-D. *Colloids Surf., A*, 316, 292 (2008).
[149] Zhao, Q., Yin, J., Feng, X., Shi, Z., Ge, Z., Jin, Z. *J. Nanosci. Nanotechnol.*, 10, 1 (2010).
[150] Yu, J.-G., Huang, K.-L., Tang, J.-C., Yang, Q., Huang, D.-S. *Int. J. Biol. Macromol.*, 44, 316 (2009).
[151] Wang, Z.-K., Hu, Q.-L., Cai, L. *Chinese J. Polym. Sci.*, 28, 801 (2010).
[152] Li, X.-B., Jiang, X.-Y. *New Carbon Mater.*, 25, 237 (2010).
[153] Lau, C., Cooney, M.J. *Langmuir*, 24, 7004 (2008).
[154] Jennings, J.A., Haggard, W.O., Bumgardner, J.D. US Patent Application, 0266694.A1 (2010).
[155] Razal, J.M., Gilmore, K.J., Wallace, G.G. *Adv. Funct. Mater.*, 18, 61 (2008).
[156] Lynam, C., Moulton, S.E., Wallace, G.G. *Adv. Mater.*, 19, 1244 (2007).
[157] Moridi mahdieh, Z., Mottaghitalab, V., Piri, N., Haghi, A.K. *Korean J. Chem. Eng.*, DOI: 10.1007/s11814-011-0129-y, *In press*

In: Development of Nanotechnology in Textiles ISBN: 978-1-62081-030-9
Editors: A. K. Haghi and G. E. Zaikov ©2012 Nova Science Publishers, Inc.

Chapter 3

CONDUCTIVE CHITOSAN/MULTIWALLED CARBON NANOTUBES (MWNTS) ELECTROSPUN NANOFIBER

V. Mottaghitalab and A. K. Haghi*
University of Guilan, Rasht, Iran

1. INTRODUCTION

Current research focuses on the electrospinning of chitosan (CHT)/multiwalled carbon nanotubes (MWNTs) composite nanofibers using a highly stable dispersion. The acetic acid (1-100%), formic acid, and trifluoroacetic acid/dichloromethane (TFA/DCM 70:30) were tested as solvents. Moreover, two different protocols were examined to establish a proper technique for the preparation of electrospinning solution. FT-IR spectroscopy utilized to infer the extent of interaction between the CHT polymer chain and MWNT filaments. A quite simple technique was employed to show the stability of electrospinning solution before the nanofiber formation process. The electrospun non-woven fabric was characterized by scanning electronic microscopic (SEM) imaging. Under optimal conditions, homogenous CHT/MWNTs nanofibers with a mean diameter of 275 nm and known physical characteristics were prepared. In the second part, a simulation algorithm has been applied for generating non-wovens with known

* Haghi@Guilan.ac.ir.

characteristics. Since the physical characteristics of simulated images are known exactly, one can employ them to test the usefulness of an algorithm for characterizing diameter and other structural features.

Over the recent decades, scientists have been interested in the creation of polymer nanofibers due to their potential in many engineering and medical properties [1]. According to various outstanding properties such as very small fiber diameters, large surface area per mass ratio, high porosity along with small pore sizes and flexibility, electrospun nanofiber mats have found numerous applications in diverse areas. For example, in the biomedical field, nanofibers play a substantional role in tissue engineering [2], drug delivery [3], and wound dressing [4]. Electrospinning is a novel and efficient method by which fibers with diameters in nanometer scale, called nanofibers, can be achieved. In the electrospinning process, a strong electric field is applied on a droplet of polymer solution (or melt) held by its surface tension at the tip of a syringe needle (or a capillary tube). As a result, the pendent drop will become highly electrified and the induced charges are distributed over its surface. Increasing the intensity of electric field, the surface of the liquid drop will be distorted to a conical shape known as the Taylor cone [5]. Once the electric field strength exceeds a threshold value, the repulsive electric force dominates the surface tension of the liquid and a stable jet emerges from the cone tip. The charged jet is then accelerated toward the target and rapidly thins and dries as a result of elongation and solvent evaporation. As the jet diameter decreases, the surface charge density increases and the resulting high repulsive forces split the jet into smaller jets. This phenomenon may take place several times leading to many small jets. Ultimately, solidification is carried out and fibers are deposited on the surface of the collector as a randomly oriented non-woven mat [6-7]. Figure 1 shows a schematic illustration of electrospinning setup.

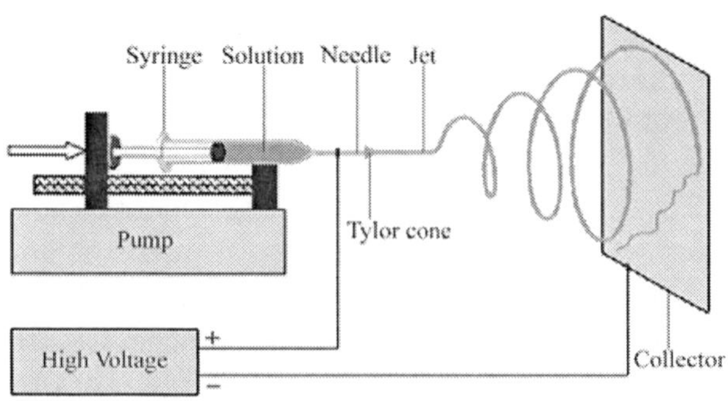

Figure 1. A typical image of the Electrospinning process [8].

The physical characteristics of electrospun nanofibers such as fiber diameter depend on various parameters which are mainly divided into three categories: solution properties (solution viscosity, solution concentration, polymer molecular weight, and surface tension), processing conditions (applied voltage, volume flow rate, spinning distance, and needle diameter), and ambient conditions (temperature, humidity, and atmosphere pressure) [9]. Numerous applications require nanofibers with desired properties suggesting the importance of the process control. This end may not be achieved unless having a comprehensive outlook of the process and quantitative study of the effects of governing parameters. In this context, Sukigara et al. [10] assessed the effect of concentration on diameter of electrospun nanofibers. They indicated that the silk nanofibers diameter increases with increasing concentration.

Beside physical characteristics, medical scientists showed a remarkable attention to biocompatiblity and biodegredablity of nanofibers made of biopolymers such as collagen [11], fibrogen [12], gelatin [13], silk [14], chitin [15] and chitosan [16]. Chitin is the second abundant natural polymer in the world and Chitosan (poly-(1-4)-2-amino-2-deoxy-β-D-glucose) is the deacetylated product of chitin [17]. CHT is well known for its biocompatible and biodegradable properties [18].

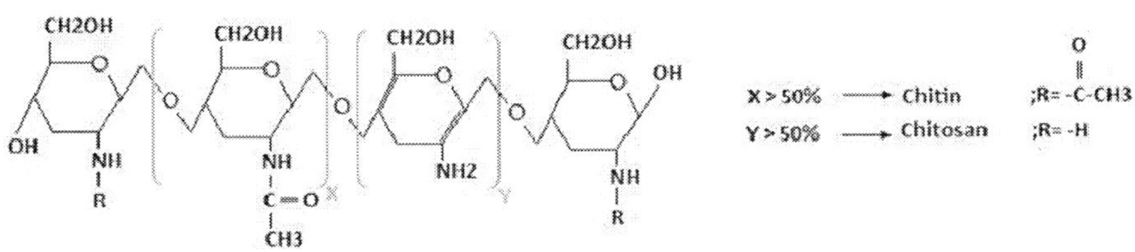

Scheme 1. Chemical structures of Chitin and Chitosan biopolymers.

Chitosan is insoluble in water, alkali, and most mineral acidic systems. However, though its solubility in inorganic acids is quite limited, chitosan is, in fact, soluble in organic acids, such as dilute aqueous acetic, formic, and lactic acids. Chitosan also has free amino groups which make it a positively-charged polyelectrolyte. This property makes chitosan solutions highly viscous and complicates its electrospinning [19]. Furthermore, the formation of strong hydrogen bonds in a 3-D network prevents the movement of polymeric chains exposed to the electrical field [20].

Different strategies have been used for bringing chitosan in nanofiber form. The three most abundant techniques includes blending of favorite polymers for electrospinning process with CHT matrix [21-22], alkali treatment of CHT backbone to improve electro spinnability through reducing viscosity [23] and employment of concentrated organic acid solution to produce nanofibers by decreasing surface tension [24]. Electrospinning of Polyethylene oxide (PEO)/CHT [21] and polyvinyl alcohol (PVA)/CHT [22] blended nanofiber are two recent studies based on the first strategy. In the second protocol, the molecular weight of chitosan decreases through alkali treatment. Solutions of the treated chitosan in aqueous 70–90% acetic acid produce nanofibers with appropriate quality and processing stability [23].

Using concentrated organic acids such as acetic acid [24] and triflouroacetic acid (TFA) with and without dichloromethane (DCM) [25-26] reported exclusively for producing neat CHT nanofibers. They similarly reported the decreasing of surface tension and at the same time enhancement of charge density of chitosan solution without significant effect on viscosity. This new method suggested significant influence of the concentrated acid solution on the reducing of the applied field required for electrospinning.

The mechanical and electrical properties of a neat CHT electrospun natural nanofiber mat can be improved by the addition of the synthetic materials including carbon nanotubes (CNTs) [27]. CNTs are one of the important synthetic polymers that were discovered by Iijima in 1991 [28]. CNTs are either single-walled nanotubes (SWNTs) or multiwalled nanotubes (MWNTs) combine together the physical properties of diamond and graphite. They are extremely thermally conductive like diamonds and appreciably electrically conductive like graphite. Moreover, the flexibility and exceptional specific surface area to mass ratio can be considered as significant properties of CNTs [29]. The scientists are becoming more interested to CNTs for the existence of exclusive properties such as superb conductivity [30] and mechanical strength for various applications. To our knowledge, there has been no report on electrospinning of CHT/MWNTs blend, except for several reports [30-31] that use PVA to improve spinnability. CNTs grafted by CHT were fabricated by the electrospinning process. In these novel sheath-core nanofibers, PVA aqueous solution has been used for enhancing nanofiber formation of MWNTs/CHT. Results showed uniform and porous morphology of the electrospun membrane. Despite adequate spinnability, total removal of PVA from nanofiber structures to form conductive substrate is not feasible. Moreover, the structural morphology and mechanical stiffness is extremely affected by thermal or alkali solution treatment of CHT/PVA/MWNTs nanofibers. The chitosan/carbon nanotube composite can be produced by the hydrogen bonds due to hydrophilic positively-charged polycation of chitosan due to amino groups and hydrophobic negatively-charged carbon nanotube due to carboxyl, and hydroxyl groups [32-34].

In a current study, it has been attempted to produce a CHT/MWNTs nanofiber without association of any type of easy electro spinnable polymers. Also, a new approach has been explored to provide highly stable and homogenous composite spinning solution of CHT/MWNTs in concentrated organic acids. This, in turn, presents homogenous conductive CHT scaffolds which is extremely important for biomedical implants.

2. EXPERIMENTAL

2.1. Materials

Chitosan polymer with degrees of deacetylation of 85% and molecular weight of 5×10^5 was supplied by Sigma-Aldrich. The MWNTs , supplied by Nutrino, have an average diameter of 4 nm and purity of about 98%. All of the other solvents and chemicals were commercially available and were used as received without further purification.

2.2. Preparation of CHT-MWNTs Dispersions

A Branson Sonifier 250 operated at 30W was used to prepare the MWNT dispersions in CHT /organic acid (90%wt acetic acid, 70/30 TFA/DCM) solution based on different protocols. In the first approach (current work) for preparation of Sample 1, the same amount (3 mg) as received MWNTs were dispersed into deionized water or DCM using solution sonicating for 10 min. A different amount of CHT was then added to MWNTs dispersion for preparation of a 8-12 wt % solution and then sonicated for another 5 min. Figure 2 shows two different protocols used in this study

In the next step, an organic acid solution was added to obtain a CHT/MWNT solution with a total volume of 5 mL and finally the dispersion was stirred for another 10 hours.Sample 2 was prepared using a second technique. The same amount of MWNTs were dispersed in chitosan solution, and the blend with a total volume of 5mL were sonicated for 10 min and dispersion was stirred for 10 hrs [35].

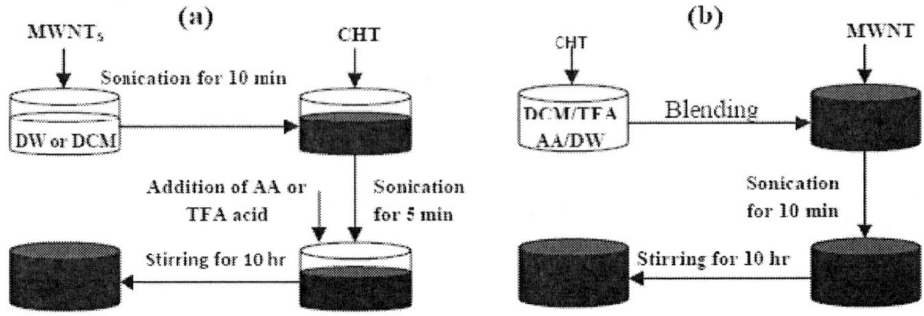

Figure 2. Two protocols used in this study for preparation of MWNTs/CHT dispersion (a) Current study(b) Ref [35].

2.2. Electrospinning of Chitosan/Carbon Nanotube Dispersion

After the preparation of spinning solution, it was transferred to a 5 *ml* syringe and became ready for the spinning of nanofibers. The experiments were carried out on a horizontal electrospinning setup shown schematically in Figure 1. The syringe containing CHT/MWNTS solution was placed on a syringe pump (New Era NE-100) used to dispense the solution at a controlled rate. A high voltage DC power supply (Gamma High Voltage ES-30) was used to generate the electric field needed for electrospinning. The positive electrode of the high voltage supply was attached to the syringe needle via an alligator clip and the grounding electrode was connected to a flat collector wrapped with aluminum foil where electrospun nanofibers were accumulated to form a non-woven mat. The voltage and the tip-to-collector distance were fixed respectively on 18-24 kV and 4-10 cm. The electrospinning was carried out at room temperature. Subsequently, the aluminum foil was removed from the collector.

2.3. Measurements and Characterizations

A small piece of mat was placed on the sample holder and gold sputter-coated (Bal-Tec). Thereafter, the micrograph of electrospun PVA fibers was obtained using scanning electron microscope (SEM, Phillips XL-30). Fourier transform infrared spectra (FTIR) were recorded using a Nicolet 560 spectrometer to investigate the interaction between CHT and MWNT in the range of 800- 4000 cm^{-1} under a transmission mode. The size distribution of the dispersed particle was evaluated with a Zetasizer (Malvern Instruments). The conductivity of the conductive fibers was measured using the four point-probe technique. A homemade four probe electrical conductivity cell operated at constant humidity has been employed. The electrodes were circular pins with a separation distance of 0.33 cm and fibers were connected to pins by silver paint (SPI). Between the two outer electrodes, a constant DC current was applied by Potentiostat/Galvanostat model 363 (Princeton Applied Research). The generated potential difference between the inner electrodes along the current flow direction was recorded by digital multimeter 34401A (Agilent). The conductivity (δ: S/cm) of the nanofiber thin film with rectangular surface can then be calculated according to length (L:cm), width (W:cm), thickness (t:cm), DC current applied (mA)and the potential drop across the two inner electrodes (mV). All measuring was repeated at least 5 times for each set of samples.

$$\delta = \frac{I \times L}{V \times W \times t} \qquad \text{Equation 1}$$

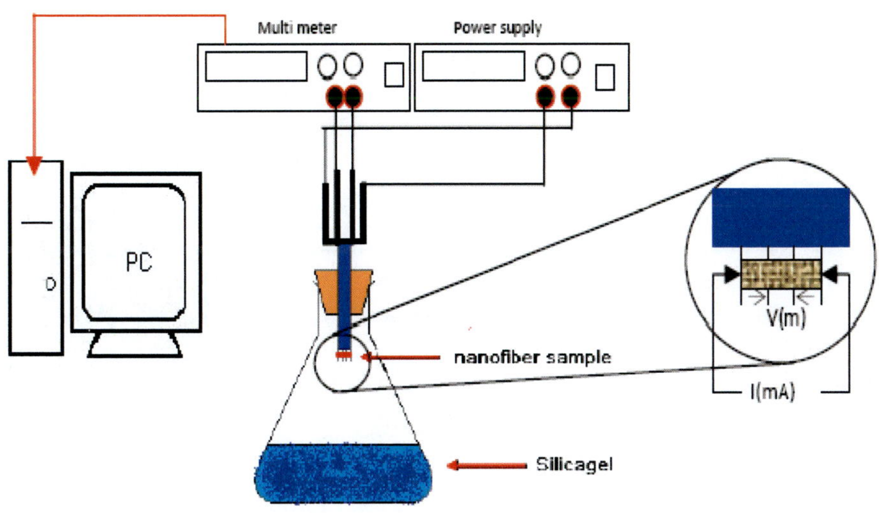

Figure 3. The experimental setup for four probe electrical conductivity measurement of nanofiber thin film.

3. RESULTS AND DISCUSSION

3.1. The Characteristics of MWNT/CHT Dispersion

Utilization of MWNTs in a biopolymer matrix initially requires their homogenous dispersion in a solvent or polymer matrix. Dynamic light scattering (DLS) is a sophisticated technique used for evaluation of particle size distribution.

DLS provides many advantages as a particle size analysis method that measures a large population of particles in a very short time period, with no manipulation of the surrounding medium. Dynamic light scattering of MWNTs dispersions indicate that the hydrodynamic diameter of the nanotube bundles is between 150 and 400 nm after 10 min of sonication for Sample 2. (Figure 4)

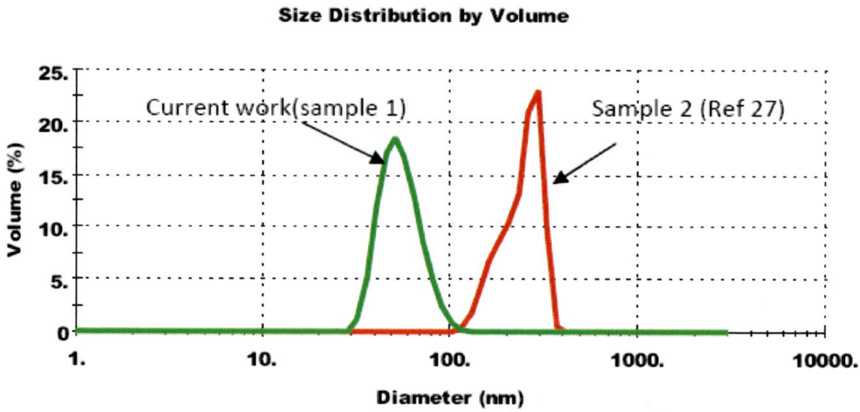

Figure 4. Hydrodynamic diameter distribution of MWNT bundles in CHT/acetic acid (1%) solution for different preparation technique.

MWNTs bundle in Sample 1 (different approach but same sonication time compared to Sample 2) shows a range of hydrodynamic diameter between 20-100 nm. (Figure 4). The lower range of hydrodynamic diameter for Sample 1 can be correlated to more exfoliated and highly stable nanotubes strands in CHT solution. The higher stability of Sample 1 compared to Sample 2 over a long period of time is confirmed by solution stability test. The results presented in Figure 3 indicate that procedure employed for preparation of sample 1 (current work) was an effective method for dispersing MWNTs in

CHT/acetic acid solution. However, MWNTs bundles in Sample 2 was found to re-agglomerate upon standing after sonication, as shown in Figure 5 where the sedimentation of large agglomerated particles is indicated.

Figure 5. Stability of CHT-MWNT dispersions (a) Current work (Sample 1)(b) Ref [35] (Sample 2).

Despite the method reported in Ref 27, neither sedimentation nor aggregation of the MWNTs bundles was observed in first sample. Presumably, this behavior in Sample 1 can be attributed to contribution of CHT biopolymer to form an effective barrier against reagglomeration of MWNTs nano particles. In fact, using sonication energy, in first step without presence of a solvent, make very tiny exfoliated but unstable particles in water as a dispersant. Instantaneous addition of acetic acid as a solvent to prepared dispersion and long mixing most likely helps the wrapping of MWNTs strands with CHT polymer chain.

Figure 6 shows the FTIR spectra of neat CHT solution and CHT/MWNTs dispersions prepared using strategies explained in the experimental part. The interaction between MWNTs and CHT in dispersed form has been understood through recognition of functional groups. The enhanced peaks at ~1600 cm^{-1} can be attributed to (N-H) band and (C=O) band of amid functional group. However the intensity of the amid group for CHT/MWNTs dispersion has been increased presumably due to a contribution of G band in MWNTs. More interestingly, in this region, the FTIR spectra of MWNTs-CHT dispersion (Sample 1) have been highly intensified compared to Sample 2 [35]. It can be correlated to a higher chemical interaction between acid functionalized C-C group of MWNTs and amid functional group in CHT.

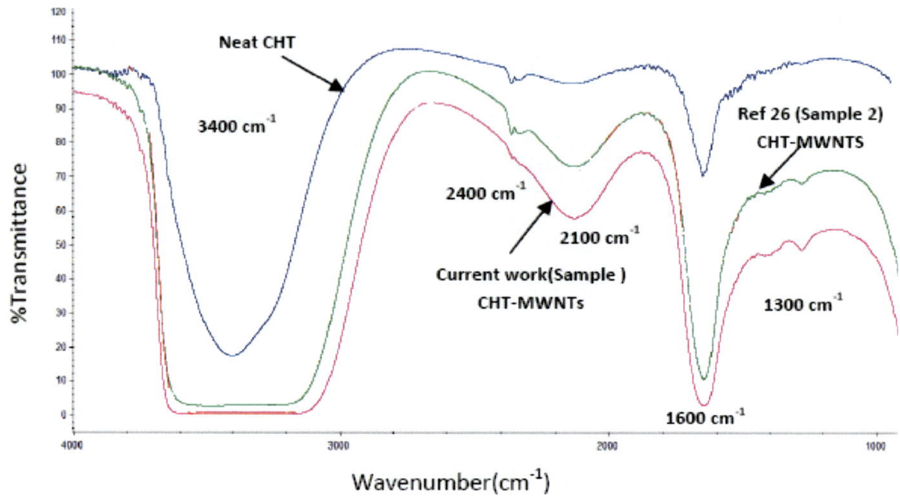

Figure 6. FTIR spectra of CHT-MWCNT in 1% acetic acid with different techniques of dispersion.

This probably is the main reason of the higher stability and lower MWNTs dimension demonstrated in Figure 4 and Figure 5. Moreover, the intensity of protonated secondary amine absorbance at 2400 cm^{-1} for a sample prepared by new technique is negligible compared to Sample 2 and neat CHT. Furthermore, the peak at 2123cm^{-1} is a characteristic band of the primary amine salt, which is associated to the interaction between positively-charged hydrogen of acetic acid and amino residues of CHT. Also, the broad peaks at ~3410 cm^{-1} due to the stretching vibration of OH group superimposed on NH stretching bond and broadens due to interhydrogen bonds of polysaccharides. The broadest peak of hydrogen bonds was observed at 3137-3588 cm^{-1} for MWNTs/CHT dispersion prepared by the new technique (Sample1).

3.2. The Physical and Morphological Characteristics of MWNTs/CHT Nanofiber

The different solvents including acetic acid 1-90%, pure formic acid, and TFA/DCM tested for the electrospinning of chitosan/carbon nanotube. No jet was seen upon applying the high voltage even above 25 kV by using of acetic acid 1-30% and formic acid as the solvent for chitosan/carbon nanotube. When the acetic acid 30-90%, used as the solvent, beads were deposited on the collector. Therefore, under these conditions, an electrospun fiber of carbon nanotube/chitosan could not be obtained (data not shown).

Figure 7 shows scanning electronic micrographs of the MWNTs/CHT electrospun nanofibers in different concentrations of CHT in TFA/DCM (70:30) solvent. As presented in Figure 7a, at low concentrations of CHT, the beads deposited on the collector and thin fibers co-exited among the beads. When the concentration of CHT was increased as shown in F,igure 7a-c the beads was decreased. Figure 7c shows homogenous electrospun nanofibers with minimum beads, thin fibers and interconnected fibers. More increasing of the concentration of CHT lead to increasing of interconnected fibers in Figure 7 d-e. Figure 8 shows the effect of concentration on average diameters of MWNTs/CHT electrospun nanofibers. Our assessments indicate that the fiber diameter of MWNTs/CHT increases with the increasing concentration. In this context, there are several studies that have reported similar to our results [36-37]. Hence, MWNTs/CHT solution in TFA/DCM (70:30) with 10 wt% chitosan resulted as optimal conditions of concentration for electrospinning. An average diameter of 275 nm (Figure 7c: diameter distribution, 148-385) investigated for this conditions. Table 1 lists the variation of nanofiber diameters and four probe electrical conductivity based on the different loading of CHT. One can expect the lower conductivity, the higher CHT content. However, this effect has been dampened by the decreasing of nanofiber diameters. This led to a nearly constant conductivity over entire measurements.

Table 1. The variation of conductivity and mean nanofiber diameter versus Chitosan loading

% CHT (%w/v)	% MWNT (%w/v)	Voltage (KV)	Tip-to-collector (cm)	Diameter (nm)	Conductivity (S/cm)
8	0.06	24	5	137 ± 58	NA
9	0.06	24	5	244 ± 61	9×10^{-5}
10	0.06	24	5	275 ± 70	9×10^{-5}
11	0.06	24	5	290 ± 87	8×10^{-5}
12	0.06	24	5	Non-uniform	NA

	magnification		Fiber diameter (nm)
	5000x	10000x	
8%	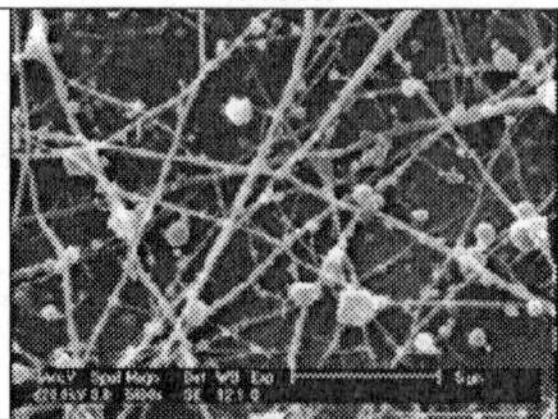	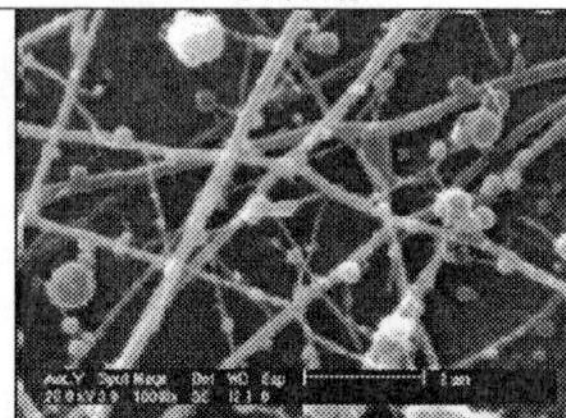	Max: 277 Min: 70 Avg: 137
9%	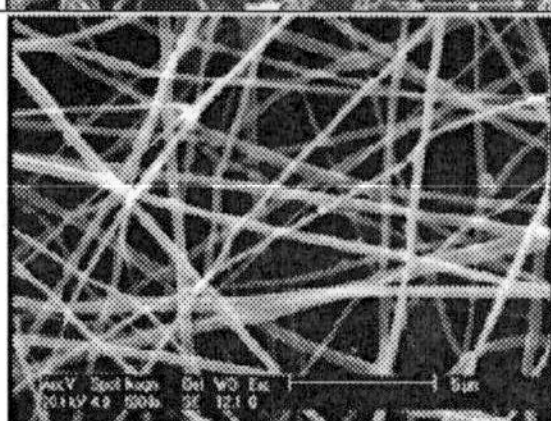	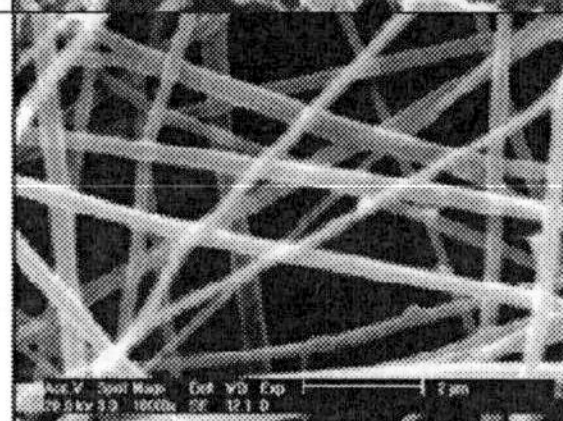	Max: 352 Min: 110 Avg: 244
10%	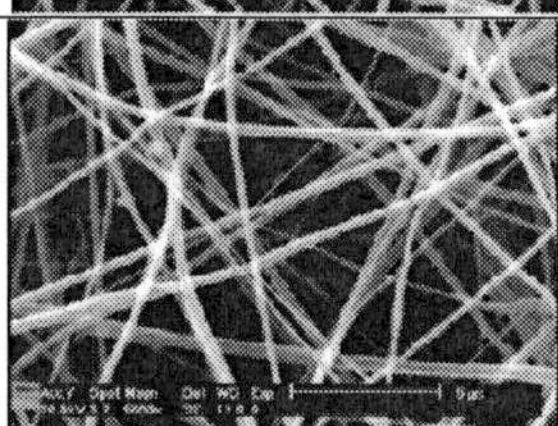	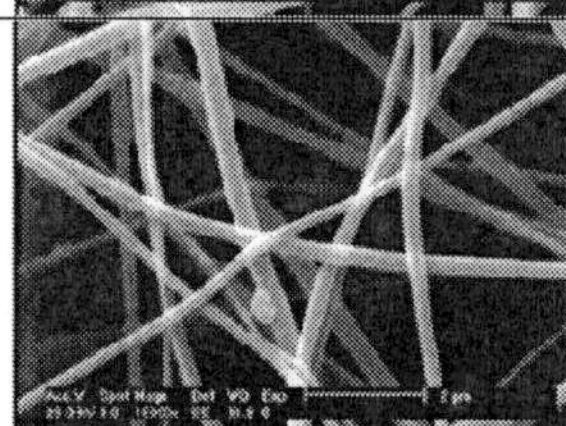	Max: 385 Min: 148 Avg: 275
11%	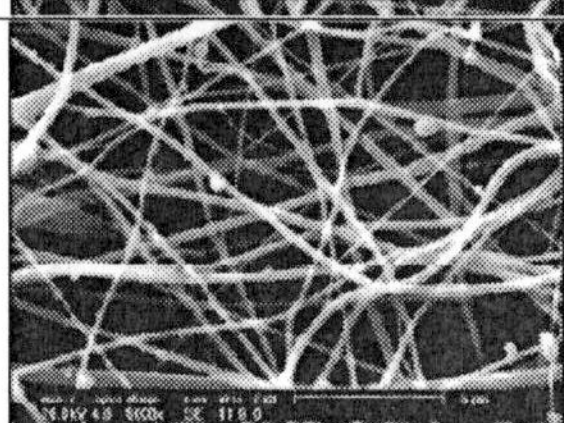	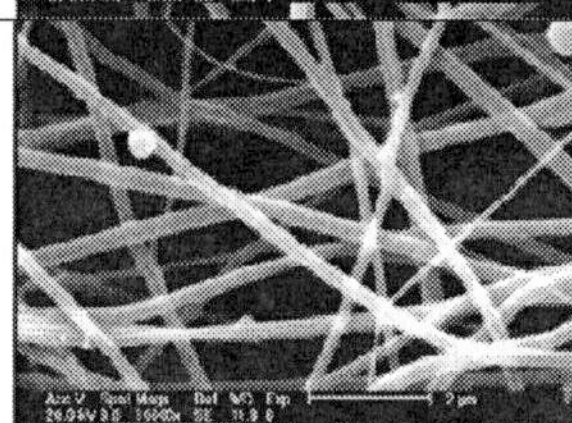	Max: 490 Min: 143 Avg: 290
12%	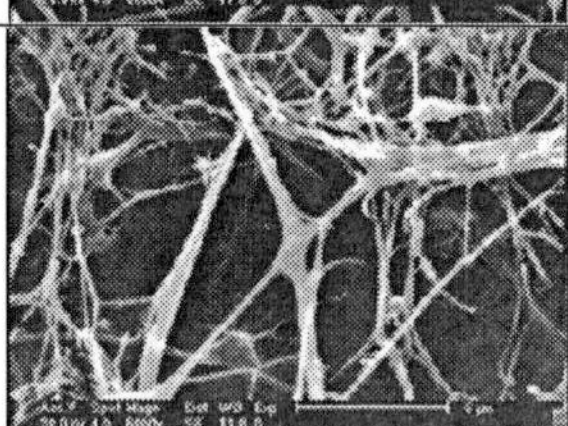	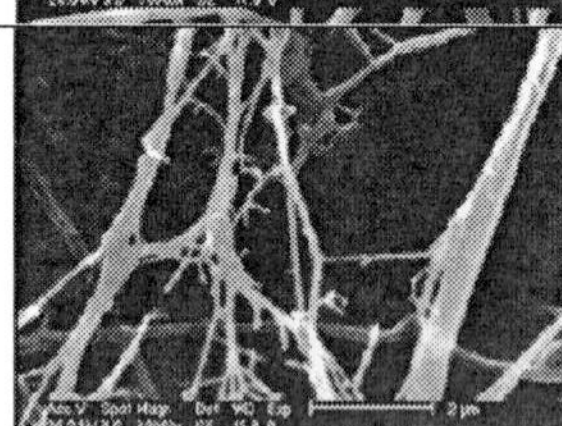	

Figure 7. Scanning electron micrographs of electrospun nanofibers at different CHT concentration (wt%): (a) 8, (b) 9, (c) 10, (d) 11, (e) 12, 24 kV, 5 cm, TFA/DCM: 70/30, (0.06%wt MWNTs).

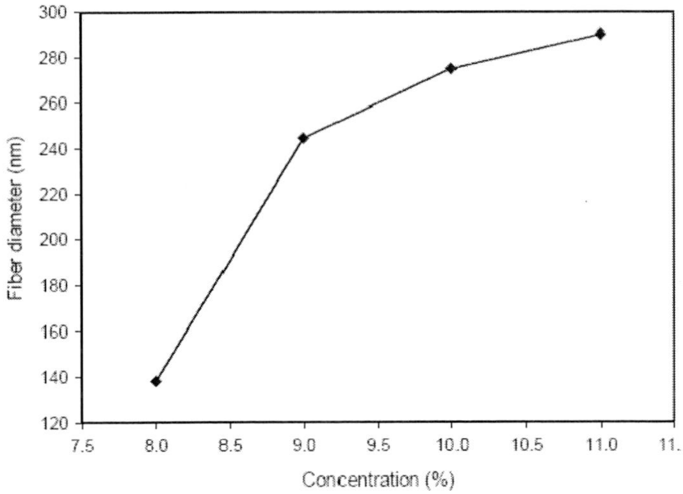

Figure 8. The effect of the concentration of CHIT/CNT dispersion on fiber diameter.

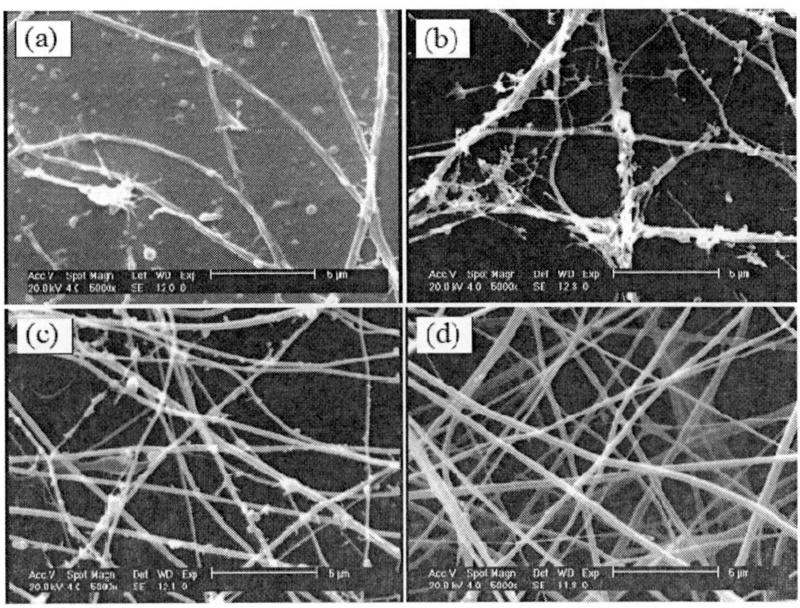

Figure 9. Scanning electronic microgarghs of electrospun fibers at different voltages (kV): (a) 18, (b) 20, (c) 22, (d) 24, 5 cm, 10 wt%, TFA/DCM: 70/30. (0.06%wt MWNTs).

To understanding the effects of voltage on morphologies of CHT/MWNT electrospun nanofibers, the SEM images in Figure 9 were analyzed. In our experiments, 18 kv were attained as threshold voltage, where fiber formation occurred. When the voltage was low, the beads and some little fiber deposited on the collector (Figure 9a). As shown in Figure 9a-d, the beads decreased by increasing voltage from 18 kV to 24 kV for electrospinning of fibers. The nanofibers collected by applying 18 kV (9a) and 20 kV (9b) were not quite clear and uniform. The higher the applied voltage, the more uniform nanofibers with less distribution starts to form. The average diameter of fibers, 22 kV (9c), and 24 kV (9d), respectively, were 204 (79-391), and 275 (148-385).

The conductivity measurement given in Table 2 confirms our observation in the first set of conductivity data. As can be seen from the last row, the amount of electrical conductivity reaches a maximum level of 9×10^{-5} at prescribed setup.

Table 2. The variation of conductivity and mean nanofiber diameter versus applied voltage

% CHT (%w/v)	% MWNT (%w/v)	Voltage (KV)	Tip-to-collector (cm)	Diameter (nm)	Conductivity (S/cm)
10	0.06	18	5	Non- uniform	NA
10	0.06	20	5	Non- uniform	NA
10	0.06	22	5	201 ± 66	6×10^{-5}
10	0.06	24	5	275 ± 70	9×10^{-5}

The distance between the tip-to-collector is another approach that controls the fiber diameter and morphology. Figure 10 shows the change in morphologies of CHT/MWNTs electrospun nanofibers at different distances between the tip-to-collector. When the distance from tip-to-collector is not long enough, the solvent could not be vaporized, hence, small amounts of interconnected thick fiber deposited on the collector (Figure 10a). In the 5 cm distance of tip-to-collector (Figure 10b), rather homogenous nanofibers have obtained with negligible beads and interconnected fibers. However, the beads increased by increasing the distance from tip-to-collector as represented from Figure 10b to Figure 10f. Similar results was observed for chitosan nanofibers fabricated by Geng et al. [24]. Also, the results show that the diameter of electrospun fibers decreased by increasing of distance tip-to-collector in Figures 10b, 10c, 10d, respectively, 275 (148-385), 170 (98-283), 132 (71-224). Similar effects of the distance between tip-to-collector on fiber diameters has observed in previous studies [38-39]. A remarkable defect and non-homogenity appears for those fibers prepared at a distance of 8 cm (Figure 10e) and 10 cm (Figure 10f). However, 5 cm for distance tip-to-collector was seen proper for electrospinning.

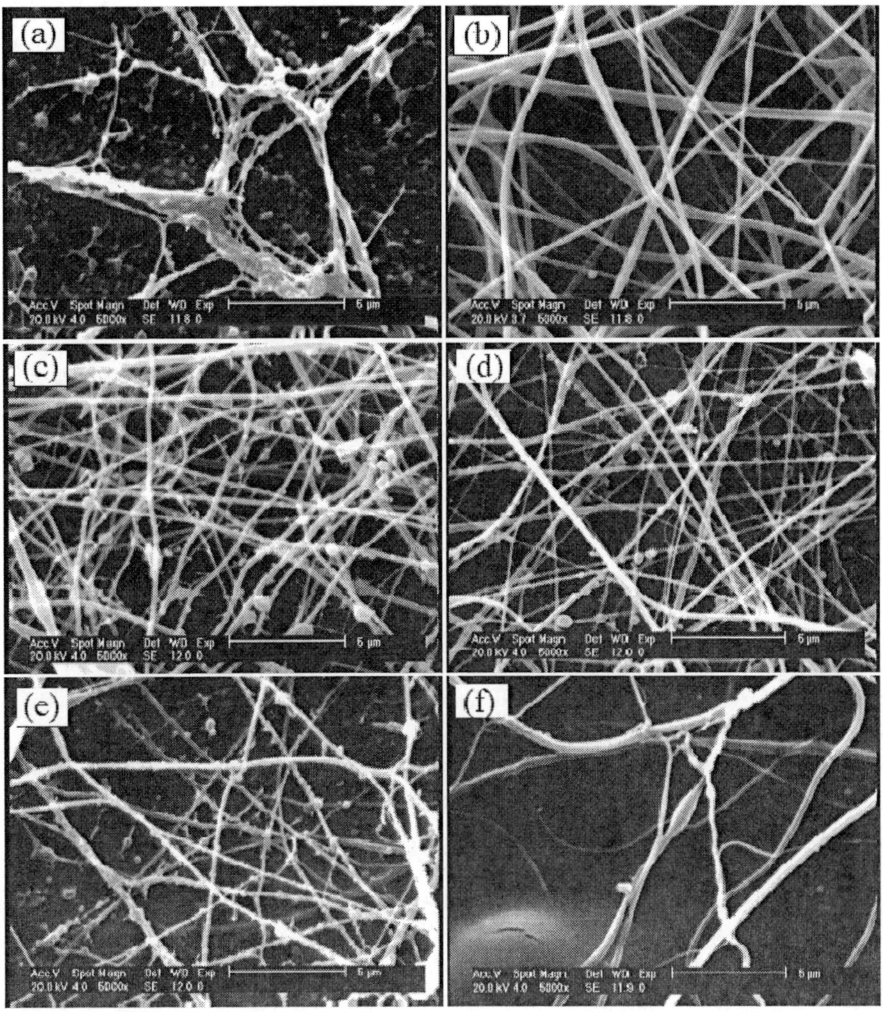

Figure 10. Scanning electronic microgragh s of electrospun fibers of Chitosan/Carbon nanotubes at different tip-to-collector distances (cm): (a) 4, (b) 5, (c) 6, (d) 7, (e) 8, (f) 10, 24 kV, 10 wt%, TFA/DCM: 70/30.

Table 3. The variation of conductivity and mean nanofiber diameter versus applied voltage

% CHT (%w/v)	% MWNT (%w/v)	Voltage (KV)	Tip-to-collector (cm)	Diameter (nm)	Conductivity (S/cm)
10	0.06	24	4	Non uniform	NA
10	0.06	24	5	275 ± 70	9×10^{-5}
10	0.06	24	6	170 ± 58	6×10^{-5}
10	0.06	24	7	132 ± 53	7×10^{-5}
10	0.06	24	8	Non uniform	NA
10	0.06	24	10	Non uniform	NA

Conductivity results also are in agreement with the data obtained in previous parts. The non-homogenity and huge bead densities plays as a barrier against electrical current and still a bead-free and thin nanofier mat shows higher conductivity compared to other samples. Experimental framework in this study was based on parameter adjusting for electro spinning of conductive CHT/MWNTs nanofiber. It can be expected that the addition of nanotubes can boost conductivity and also change morphological aspects which is extremely important for biomedical applications.

Conclusion

Conductive composite nanofibers of CHT/MWNTs hasve been produced using the conventional electrospinning technique. A new protocol suggested for prepararation of electrospinning soilution which shows much better stability and homogeneity compared to previous techniques. Several solvents including acetic acid 1-90%, formic acid, and TFA/DCM (70:30) were investigated in the electrospinning of CHT/MWNTs dispersion. It was observed that the TFA/DCM (70:30) solvent is most preferred for fiber formation process with acceptable electrospinnability. The formation of nanofibers with conductive pathways regarding to exfoliated and interconntected nanotube strands is a breakthrough in chitosan nanocomposite area. This can be considered as a significant improvement in electrospinning of chitosan/carbon nanotube dispersion. It has also been observed that the homogenous fibers with an average diameter of 275 nm could be prepared with a conductivity of 9×10^{-5}.

Appendix A

A.1. Basic Factors in Nanoscale Systems

In addition to size effects that occur in micro-and nanoflyuidike, we should consider the following basic factors:

- Surface roughness;
- Dissolved gases;
- Chemical surface properties;
- Hydrophobic - hydrophilic of surface;
- Contaminants;
- Heating due to uncontrollable processes;
- Electrical properties of the surface.

Viscous forces in the fluid can lead to large dispersion flow along the axis of motion. They have a significant impact, both on the scale of individual molecules, and the scale of microflows - near the borders of the liquid-solid (beyond a few molecular layers), during the motion on complex and heterogeneous borders.

Influence of the effect of boundary regions on the particles and fluxes observed experimentally in the range of molecular thicknesses up to hundreds of nanometers. If the surface has superhydrophobic properties, this range can extend to the micron thickness. *Molecular theory can predict the effect of hydrophobic surfaces in the system only up to tens of nanometers.*

Fluids, the flow of liquid or gas, have properties that vary continuously under the action of external forces. In the presence of fluid, shear forces are small in magnitude, leading to large changes in the relative position of the element of fluid. In contrast, changes in the relative positions of atoms in solids remain small under the action of any small external force. Termination of action of the external forces on the fluid does not necessarily lead to the restoration of its initial form.

Capillary Effects

To observe the capillary effects, you must open the nanotube, that is, remove the upper part, or lids. Fortunately, this operation is quite simple.

The first study of capillary phenomena has shown that there is a relationship between the magnitude of surface tension and the possibility of its being drawn into the channel of the nanotube. It was found that the liquid penetrates into the channel of the nanotube, if its surface tension is not higher than 200 mN / m. For example, to enter into the channel of the nanotube of certain metals, concentrated nitric acid is used, the surface tension is low (43 mN/m). Then spend a joke at 4000S for 4 h in an atmosphere of hydrogen, which leads to the recovery of the metal. Thus, the obtained nanotubes containing nickel, cobalt and iron.

The metal-walled carbon nanotubes can be filled with gaseous substances, such as hydrogen in molecular form. This ability is of great practical importance, opening the ability to safely store hydrogen, which can be used as a clean fuel in internal combustion engines.

Specific Electrical Resistance of Carbon Nanotubes

Due to the small size of carbon nanotubes, only in 1996 was there success in directly measuring their electrical resistivity ρ. The results of direct measurements showed that the resistivity of the nanotubes can be varied within wide limits to 0.8 ohm / cm. The minimum value is lower than that of graphite. Most of the nanotubes have metallic conductivity, and the smaller shows properties of a semiconductor with a bandgap of 0.1 to 0.3 eV.

The resistance of a single-walled nanotube is independent of its length, because of this, it is convenient to use for the connection of logic elements in microelectronic devices. The permissible current density in carbon nanotubes is much greater than in metallic wires of the same cross-section and one hundred times better achievement for superconductors.

Emission Properties of Carbon Nanotubes

The results of the study of emission properties of the material, where the nanotubes were oriented perpendicular to the substrate, have been very interesting for practical use. An attained value of the emission current density is in the order of 0.5 mA / mm^2 .

A.2 Electrokinetic Processes in Micro - and Nanoscale Systems

The most effective and common way to control microflow substances are *electrokinetic* and *hydraulic*. At the same time, the most technologically advanced and automated are considered electrokinetic.

Charges transfer in mixtures occurs as a result of the directed motion of charge carriers - ions. There are different mechanisms of such transfer, but usually they are *convection, migration and diffusion*.

Convection is called mass transfer of the macroscopic flow. *Migration* - the movement of charged particles by electrostatic fields. The velocity of the ions depends on field strength. In microfluidics, a special role is played in the *electrokinetic processes* that can be divided into four types: *electro-osmosis, electrophoresis, streaming potential and sedimentation potential.* These processes can be qualitatively described as follows:

a) *electro-osmosis* - the movement of the fluid volume in response to the applied electric field in the channel of the electrical double layers on its wetted surfaces.
b) *Electrophoresis* - the forced motion of charged particles or molecules, in mixture with the acting electric field.
c) Streamy potential - the electric potential, which is distributed through a channel with charged walls, in the case when the fluid moves under the action of pressure forces. Joule electric current associated with the effect of charge transfer is flowing stream.
d) The potential of sedimentation - an electric potential is created when charged particles are in motion relative to a constant fluid.

In general, the microchannel cross-section amount of introduced probe (when entering electrokinetic method) depends on the applied voltage U , time during which the received power, and mobility of the sample components μ :

$$Q = \frac{\mu SUt}{L} \cdot c$$

Where

c – Probe concentration in the mixture,

L – channel length

Amount of injected substance is determined by the electrophoretic and total electro-osmotic mobilities μ .

In the hydrodynamic mode of entry by the pressure difference in the channel or capillary of circular cross-section, the volume of injected probe V_c :

$$V_c = \frac{4}{128} \cdot \frac{\Delta p \pi dt}{\eta L}$$

Where

Δp - Pressure differential,

d - Diameter of the channel,

η - Viscosity.

A.3. Continuum Hypothesis

In the simulation of processes in micron-fundamental systems are the following basic principles:

1) Hypothesis of *laminar* flow (sometimes considered taken for granted when it comes to microfluidics);
2) Continuum hypothesis (detection limits of applicability);
3) Laws of formation of the velocity profile, mass transfer, the distribution of electric and thermal fields;
4) Boundary conditions associated with the geometry of structural elements (walls of channels, mixers zone flows, etc.).

Since we consider the physical and chemical transport processes of matter and energy, mathematical models, most of them have the form of systems of differential equations of second order partial derivatives. Methods for solving such equations - analytical (Fourier and its modifications, such as the method of Greenberg, Galerkin, in some cases, the method of d'Alembert and the Green's functions, the Laplace operator method, etc.) or numerical (explicit or, more effectively, implicit finite difference schemes) - traditional. The development involves, basically, numerical methods and follows the path of saving computing resources, and increasing the speed of modern computers.

The small size of the microchannels (typical dimensions of 5 to 300 microns) and low surface roughness create good conditions for the establishment of laminar flow. Traditionally, the image of the nature of the flow gives the dimensionless characteristic numbers: the Reynolds number and friction factor Darcy.

In the motion of fluids in channels, the turbulent regime is rarely achieved.. At the same time, the movement of gases is usually turbulent.

Although the liquid - quantized in the length scale of intermolecular distances (about 0.3 nm to 3 nm in liquids and for gases), they are assumed to be continuous in most cases, microfluidics. Continuum hypothesis (continuity, continuum) suggests that the macroscopic properties of fluids consisting of molecules are the same as if the fluid were completely continuous (structurally homogeneous). Physical characteristics: mass, momentum and energy associated with the volume of fluid containing a sufficiently large number of molecules must be taken as the sum of all of the relevant characteristics of the molecules.

Continuum hypothesis leads to the concept of fluid particles. In contrast to the ideal of a point particle in ordinary mechanics, fluid mechanics, particle in the fluid has a finite size.

At the atomic scale, we would see large fluctuations due to the molecular structure of fluids, but with the increase in sample size, we reach a level where it is possible to obtain stable measurements. This volume of probe must contain a sufficiently large number of molecules to obtain reliable reproducible signals with small statistical fluctuations. For example, if we determine the required volume as a cube with sides 10 nm, this volume contains some of the molecules and determines the level of fluctuations of the order of 0.5%.

The most important provision requiring verification, a valid analysis of mass transfer of substances based on a continuum model that can be used instead of the statistical analysis ensemble of individual particles is the concentration dependence. The position of the continuum model is considered a necessary condition for microfluidics.

The applicability of the hypothesis is based on comparison of the free path length of a particle in a liquid with a characteristic geometric size d . The ratio of these lengths - the Knudsen number: $Kn = \lambda / d$. Based on estimates of the Knudsen number, there are two important statements:

a) $Kn < 10^{-3}$ - justified hypothesis of a continuous medium,

b) $Kn < 10^{-1}$ - allowed the use of adhesion of particles to the solid walls of the channel.

Wording of the last condition can also be varied: both in form and in a more complex form, associated with shear stresses. The calculation of λ can be carried out as $\lambda \approx \sqrt[3]{V / Na}$,

Where

$\overline{V}$ - molar volume,

Na – Avogadro's number.

Under certain geometrical approximations of the particles of substance, free path length can be calculated as $\lambda \approx 1/\left(\sqrt{2}\pi r_S^2 Na\right)$, if used instead of the Stokes radius, as a consequence of the spherical approximation of the particle. On the other hand, for a rigid model of the molecule, it should be replaced by the characteristic size of the particles - the radius of inertia, calculated as $R_g = n_i \cdot \delta_l / \sqrt{6}$.

Here, the length of a fragment of the chain (link), - the number of links.

Of course, the hypothesis of a continuous medium is not acceptable when the system in question approaches the molecular scale. This happens in a nanoliquid, such as liquid transport through nano-pores in cell membranes or artificially-made nanochannels.

A.4. The Molecular Dynamics Method

In contrast to the continuum hypothesis, the essence of modeling the molecular dynamics method is as follows. We consider a large ensemble of particles which simulate atoms or molecules, i.e., all atoms are material points. It is believed that the particles interact with each other and, moreover, may be subject to external influence. Inter-atomic forces are represented in the form of the classical potential force (the gradient of the potential energy of the system).

The interaction between atoms is described by means of van der Waals forces (intermolecular forces), mathematically expressed by the Lennard-Jones potential:

$$V(r) = \frac{Ae^{-\sigma r}}{r} - \frac{C_6}{r^6}$$

Where

A and - some coefficients depending on the structure of the atom or molecule, σ - the smallest possible distance between the molecules.

In the case of two isolated molecules at a distance of the interaction force is zero, the repulsive forces balance attractive forces. When the resultant force is the force of gravity, which increases in magnitude, reaching a maximum and then decreases. This is when a repulsive force should be considered. Molecules in the field of these forces has potential energy $V(r)$, which is connected with the force of by the differential equation

$$dV = -f(r)dr$$

At the point $r = r_0$, $f(r) = 0$, reaches an extremum (minimum).

The chart of such a potential is shown below in Figure A.1. The upper (positive) half-axis corresponds to the repulsion of the molecules, the lower (negative) half-plane - their attraction.

We can say simply that at short distances, the molecules repel each other mainly on the long - drawn. Based on this hypothesis, and now an obvious fact, the van der Waals and received its equation of the state for real gases.

The exponential summand in the expression for the potential describing the repulsion of the molecules at small distances, often approximated as

$$\frac{Ae^{-\sigma r}}{r} \approx \frac{C_{12}}{r^{12}}$$

In this case, we obtain the Lennard-Jones potential:

$$V(r) = \frac{C_{12}}{r^{12}} - \frac{C_6}{r^6} \qquad \text{(A.1)}$$

The interaction between carbon atoms is described by the potential

$$V_{CC}(r) = K(r-b)^2,$$

where - constant tension (compression) connection, - the equilibrium length of connection, - current length of the connection.

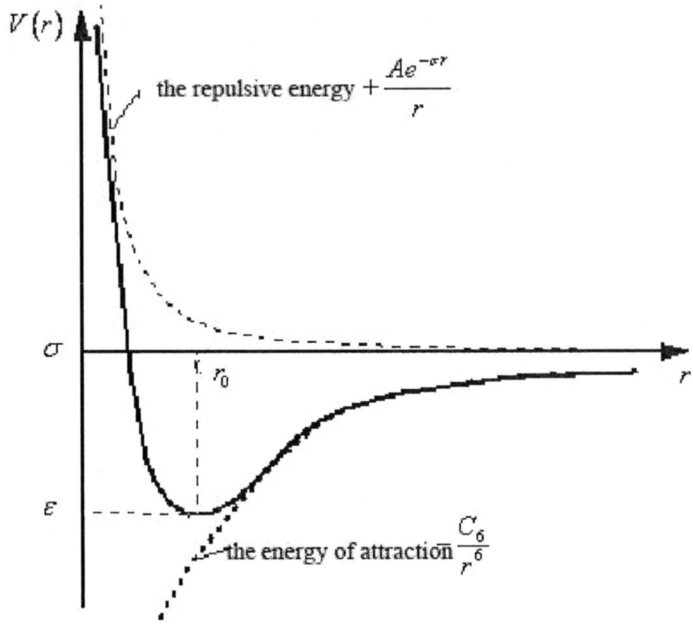

Figure A.1. The chart of the potential energy of intermolecular interaction.

The interaction between the carbon atom and hydrogen molecule is described by the Lennard-Jones potential

$$V(r) = 4\varepsilon\left[\left(\frac{\sigma}{r}\right)^{12} - \left(\frac{\sigma}{r}\right)^{6}\right]$$

with parameters $\varepsilon = 2{,}76 мэВ = 32 кДж$, $\sigma = 3{,}17 A$.

For all particles (Figure A.2), the equations of motion are written:

$$m\frac{d^2\overline{r_i}}{dt^2} = \overline{F}_{T-H_2}(\overline{r_i}) + \sum_{j \neq i}\overline{F}_{H_2-H_2}\left(\overline{r_i} - \overline{r}_j\right),$$

where - force, acting by the CNT, - force acting on the i -th molecule from the j -th molecule

The coordinates of the molecules are distributed regularly in the space, the velocity of the molecules are distributed according to the Maxwell equilibrium distribution function according to the temperature of the system:

$$f(u,v,w)=\frac{\beta^3}{\pi^{3/2}}\exp\left(-\beta^2\left(u^2+v^2+w^2\right)\right)\quad \beta=\frac{1}{\sqrt{2RT}}$$

The macroscopic flow parameters are calculated from the distribution of positions and velocities of the molecules:

$$\overline{V}=\left\langle \overline{v}_i\right\rangle=\frac{1}{n}\sum_i \overline{v}_i,\qquad \rho=\frac{nm}{V_0},\qquad \frac{3}{2}RT=\frac{1}{2}\left\langle\left|\overline{v}_i^{/}\right|^2\right\rangle,\quad \overline{v_i^{/}}=\overline{v}_i-\overline{V},$$

The resulting system of equations is solved numerically. However, the molecular dynamics method has limitations of applicability: 1) the de Broglie wavelength (where - Planck's constant, - the mass of the particle, - velocity) of the atom must be much smaller than the inter-atomic distance;

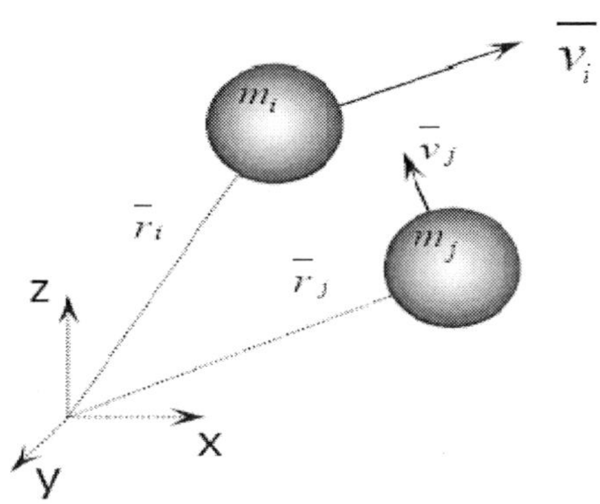

Figure A.2.

2) Classic molecular dynamics can not be applied for modeling systems consisting of light atoms such as helium or hydrogen; 3) at low temperatures, quantum effects become decisive for the consideration of such systems must

use quantum chemical methods; 4) necessary that the time at which we consider the behavior of the system were more than the relaxation time of the physical quantities.

A.5. van Der Waals Equation

In 1873, van der Waals proposed an equation of state is a qualitatively good description of liquid and gaseous systems. It is for one mole (one mole) is:

$$\left(p+\frac{a}{v^2}\right)(v-b)=RT \qquad \text{(A.2)}$$

Note that at this equation becomes the equation of the state of ideal gas

$$pv = RT \qquad \text{(A.3)}$$

van der Waals equation can be obtained from the Clapeyron equation of Mendeleev by an amendment to the magnitude of the pressure and the amendment to the volume, both constant and independent of and but dependent on the nature of the gas.

The amendment takes into account:

1) The volume occupied by the molecules of real gas (in an ideal gas molecules are taken as material points, not occupying any volume);

2) So-called "dead space", which cannot penetrate the molecules of real gas during motion, i.e. volume of gaps between the molecules in their dense packing.

Thus, (Figure A.3). The amendment takes into account the interaction force between the molecules of real gases. It is the internal pressure, which is determined from the following simple considerations.

Two adjacent elements of the gas will react with a force proportional to the product of the quantities of substances enclosed in these elementary volumes.

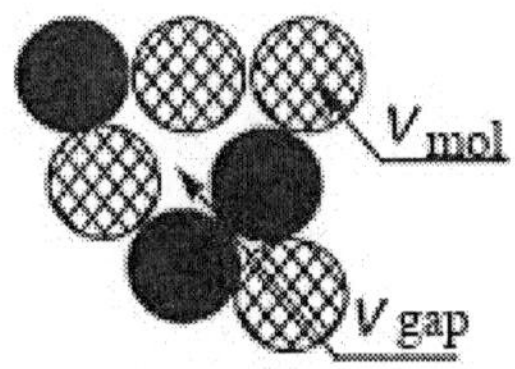

Figure A.3. Location scheme of molecules in a real gas.

Therefore, the internal pressure is proportional to the square of the concentration n :

$$p_{вн} \sim n^2 \sim \rho^2 \sim \frac{1}{v^2},$$

Where

ρ - the gas density.

Thus, the total pressure consists of internal and external pressures:

$$p + p_{вн} = p + \frac{a}{v^2}$$

Equation (A.3) is the most common for an ideal gas. Under normal physical conditions ($p_0 = 0{,}1013МПа, t_0 = 0^0 C$)

$v\mu = 22{,}4м^3 /(кмоль \cdot K^0)$, and then from (A.3) we obtain:

$$R\mu = \frac{pv\mu}{T} = \frac{0{,}1013 \cdot 10^6 \cdot 22{,}4}{273} = 8314 \frac{дж}{кмоль \cdot K^0}$$

Knowing that you can find the gas constant for any gas by help of the value of its molecular mass (Table A.1):

$$R = \frac{R\mu}{\mu} = \frac{8314}{\mu}$$

Table A.1. The molecular weight ($[\mu] = кг/кмоль$) of some gases

Gas	N	Ar	H_2	O_2	CO	CO_2	ammonia	air
μ	28	40	2	32	28	44	17	29

For gas mixtures with mass state, the equation has the form:

$$pv = MR_{см}T = \frac{8314MT}{\mu_{см}} \tag{A.4}$$

Where

$R_{см}$ - gas constant of the mixture.

The gas mixture can be given by the mass proportions g_i, voluminous or mole fractions, respectively, which are defined as the ratio of mass m_i, volume or number of moles of gas to total mass M, volume or number of moles of gas mixture. Mass fraction of component is $g_i = \frac{m_i}{M}$, where $i = 1, n$. It is obvious that and $\sum_{i=1}^{n} g_i = 1$. The volume fraction is $r_i = \frac{v_i}{v_{см}}$, where - partial volume of component mixtures.

Similarly, we have $\sum_{i=1}^{n} v_i = v_{см}, \sum_{i=1}^{n} r_i = 1$.

Depending on specificity of tasks, the gas constant of the mixture may be determined as follows:

$$R_{см} = \sum_{i=1}^{n} g_i R_i; \qquad R_{см} = \frac{1}{\sum_{i=1}^{n} r_i R_i^{-1}}$$

If we know the gas constant R_{CM}, the seeming molecular weight of the mixture is equal to

$$\mu_{CM} = \frac{8314}{R_{CM}} = \frac{8314}{\sum_{i=1}^{n} g_i R_i} = 8314 \sum_{i=1}^{n} r_i R_i^{-1}$$

The pressure of the gas mixture is equal to the sum of the partial pressures of individual components in the mixture p_i :

$$p = \sum_{i=1}^{n} p_i \tag{A.5}$$

Partial pressure - pressure that has gas, if it is one at the same temperature fills the whole volume of the mixture ($p_i v_{CM} = RT$).

With various methods of setting the gas mixture, partial pressures

$$p_i = pr_i;$$

$$p_i = \frac{pg_i \mu_{CM}}{\mu_i} \tag{A.6}$$

From the expression (A.6), we see that for the calculation of the partial pressures, itis necessary to know the pressure of the gas mixture, the volume or mass fraction of the gas component, as well as the molecular weight of the gas mixture and the molecular weight of of gas μ_i .

The relationship between mass and volume fractions are written as follows:

$$g_i = \frac{m_i}{m_{CM}} = \frac{\rho_i v_i}{\rho_{CM} v_{CM}} = \frac{R_{CM}}{R_i} r_i = \frac{\mu_i}{\mu_{CM}} r_i$$

We rewrite (A.2) as

$$v^3 - \left(b + \frac{RT}{p}\right)v^2 + \frac{a}{p}v - \frac{ab}{p} = 0. \quad \text{(A.7)}$$

When and $T = T_k$, where and - critical pressure and temperature, all three roots of (A.7) are equal to the critical volume

$$v^3 - \left(b + \frac{RT_k}{P_k}\right)v^2 + \frac{a}{p_k}v - \frac{ab}{p_k} = 0. \quad \text{(A.8)}$$

Because $v_1 = v_2 = v_3 = v_k$, then equation (A.8) must be identical to the equation

$$(v - v_1)(v - v_2)(v - v_3) = (v - v_k)^3 = v^3 - 3v^2 v_k + 3vv_k^2 - v_k^3 = 0. \quad \text{(A.9)}$$

Comparing the coefficients at the equal powers of in both equations leads to the equalities

$$b + \frac{RT_k}{p_k} = 3v_k; \quad \frac{a}{p_k} = 3v_k^2; \quad \frac{ab}{p_k} = v_k^3. \quad \text{(A.10)}$$

Hence

$$a = 3v_k^2 p_k;$$

$$b = \frac{v_k}{3}. \quad \text{(A.11)}$$

Considering (A.10) as equations for the unknowns p_k, v_k, T_k, we obtain

$$p_k = \frac{a}{27b^2};$$

$$v_k = 3b;$$

$$T_k = \frac{8a}{27bR}. \quad (A.12)$$

From (A.10) and (A.11) or (A.12) we can find the relation

$$\frac{RT_k}{p_k v_k} = \frac{8}{3} \quad (A.13)$$

Instead of the variables p, v, let us introduce the relationship of these variables to their critical values (leaden dimensionless parameters)

$$\pi = \frac{p}{p_k};\ \omega = \frac{v}{V_k};\ \tau = \frac{T}{T_k}. \quad (A.14)$$

Substituting (A.12) and (A.14) in (A.7) and using (A.13), we obtain

$$\left(\pi p_k + \frac{3v_k^2 p_k}{\omega^2 v_k^2}\right)\left(\omega v_k - \frac{v_k}{3}\right) = RT_k\tau,$$

$$\left(\pi + \frac{3}{\omega^2}\right)(3\omega - 1) = 3\frac{RT_k}{p_k v_k}\tau,$$

$$\left(\pi + \frac{3}{\omega^2}\right)(3\omega - 1) = 8\tau \quad (A.15)$$

In (A.15), are not permanent and b, depending on the nature of the gas. That is, if the units of measurement of pressure, volume and temperature to use their critical values (use the leaden parameters), the equation of state is the same for all substances.

This conditionis called the *law of corresponding states.*

REFERENCES

[1] Agarwal, S., Wendorff, J.H., Greiner, A. *Use of electrospinning technique for biomedical applications.* Polymer, 2008. 49(26): p. 5603-5621.

[2] Li, M., et al., *Electrospun protein fibers as matrices for tissue engineering.* Biomaterials, 2005. 26(30): p. 5999-6008.

[3] Zeng, J., et al., Influence of the drug compatibility with polymer solution on the release kinetics of electrospun fiber formulation. Journal of Controlled Release, 2005. 105(1-2): p. 43-51.

[4] Khil, M.-S., et al., *Electrospun nanofibrous polyurethane membrane as wound dressing.* Journal of Biomedical Materials Research Part B: Applied Biomaterials, 2003. 67B(2): p. 675-679.

[5] Taylor, G.I. *Electrically driven jets.* Proc Roy Soc London 1969. 313: p. 453-475.

[6] Doshi, J. Reneker, D.H. *Electrospinning process and applications of electrospun fibers.* Journal of Electrostatics, 1995. 35(2-3): p. 151-160.

[7] Li, D. Xia, Y. *Electrospinning of Nanofibers: Reinventing the Wheel?* Advanced Materials, 2004. 16(14): p. 1151-1170.

[8] Ziabari, M., Mottaghitalab, V., Haghi, A.K. *Evaluation of electrospun nanofiber pore structure parameters.* Korean J. Chem. Eng., 2008. 25(4): p. 923-932.

[9] Tan, S.H., et al., Systematic parameter study for ultra-fine fiber fabrication via electrospinning process. Polymer, 2005. 46(16): p. 6128-6134.

[10] Sukigara, S., et al., Regeneration of Bombyx mori silk by electrospinning—part 1: processing parameters and geometric properties. Polymer, 2003. 44: p. 5721-5727.

[11] Matthews, J.A., et al., *Electrospinning of collagen nanofibers.* Biomacromolecules, 2002. 3(2): p. 232-8.

[12] McManus, M.C., et al., Electrospun fibrinogen: Feasibility as a tissue engineering scaffold in a rat cell culture model. Journal of Biomedical Materials Research Part A, 2007. 81A(2): p. 299-309.

[13] Huang, Z.-M., et al., Electrospinning and mechanical characterization of gelatin nanofibers. Polymer, 2004. 45(15): p. 5361-5368.

[14] Zhang, X., Reagan, M.R., Kaplan, D.L. *Electrospun silk biomaterial scaffolds for regenerative medicine.* Advanced Drug Delivery Reviews, 2009. 61(12): p. 988-1006.

[15] Noh, H.K., et al., Electrospinning of chitin nanofibers: Degradation behavior and cellular response to normal human keratinocytes and fibroblasts. Biomaterials, 2006. 27(21): p. 3934-3944.

[16] K. Ohkawa, et al., *Chitosan Nanofiber.* Biomacromolecules, 2006. 7: p. 3291-3294.

[17] Agboh, O.C. Qin, Y. *Chitin and Chitosan Fibers.* Polymers for Advanced Technologies, 1997. 8: p. 355-365.

[18] Rinaudo, M., *Chitin and chitosan: Properties and applications.* Prog. Polym. Sci., 2006. 31: p. 603-632.

[19] Aranaz, I., et al., *Functional Characterization of Chitin and Chitosan.* Current Chemical Biology, 2009. 3: p. 203-230.

[20] Neamnark, A., Rujiravanit, R., Supaphol, P. *Electrospinning of hexanoyl chitosan.* Carbohydrate Polymers, 2006. 66(3): p. 298-305.

[21] Duan, B., et al., Electrospinning of chitosan solutions in acetic acid with poly(ethylene oxide). Vol. 15. 2004, Leiden, PAYS-BAS: Brill. 15.

[22] Jia, Y.T., et al., Fabrication and characterization of poly (vinyl alcohol)/chitosan blend nanoWbers produced by electrospinning method. Carbohydrate Polymers, 2007. 67: p. 403-409.

[23] Homayoni, H., Ravandi, S.A.H., Valizadeh, M. *Electrospinning of chitosan nanofibers: Processing optimization.* Carbohydrate Polymers, 2009. 77: p. 656-661.

[24] Geng, X., Kwon, O.H., Jang, J. *Electrospinning of chitosan dissolved in concentrated acetic acid solution.* Biomaterials, 2005. 26: p. 5427-5432.

[25] TORRES-GINER, et al., Development of Active Antimicrobial Fiber-Based Chitosan Polysaccharide Nanostructures using Electrospinning. 2008, Weinheim, ALLEMAGNE: Wiley. 12.

[26] Vrieze, S.D., et al., *Electrospinning of chitosan nanofibrous structures: feasibility study.* Journal of Materials Science, 2007. 42(19): p. 8029-8034.

[27] Ohkawa, K., et al., *Electrospinning of Chitosan.* Macromolecular Rapid Communications, 2004. 25(18): p. 1600-1605.

[28] Iijima, S., *Helical microtubules of graphitic carbon.* Nature, 1991. 354(6348): p. 56–58.

[29] Esawi, A.M.K. Farag, M.M. *Carbon nanotube reinforced composites: Potential and current challenges.* Materials & Design, 2007. 28(9): p. 2394-2401.

[30] Feng, W., et al., The fabrication and electrochemical properties of electrospun nanofibers of a multiwalled carbon nanotube grafted by chitosan. Nanotechnology, 2008. 19(10): p. 105707.

[31] Liao, H., et al., Improved cellular response on multiwalled carbon nanotube-incorporated electrospun polyvinyl alcohol/chitosan nanofibrous scaffolds. Colloids and Surfaces B: Biointerfaces, 2011. In Press, Accepted Manuscript(doi:10.1016/j.colsurfb.2011.02.010): p. 1-28.

[32] Baek, S.-H., Kim, B., Suh, K.D. *Chitosan particle/multiwall carbon nanotube composites by electrostatic interactions.* Colloids and Surfaces A: Physicochemical and Engineering Aspects, 2008. 316(1-3): p. 292-296.

[33] Liu, Y.-L., Chen, W.H., Chang, Y.H. Preparation and properties of chitosan/carbon nanotube nanocomposites using poly(styrene sulfonic acid)-modified CNTs. Carbohydrate Polymers, 2009. 76(2): p. 232-238.

[34] Tkac, J., Whittaker, J.W., Ruzgas, T. The use of single walled carbon nanotubes dispersed in a chitosan matrix for preparation of a galactose biosensor. Biosensors and Bioelectronics, 2007. 22(8): p. 1820-1824.

[35] Spinks, G.M., et al., *Mechanical properties of chitosan/CNT microfibers obtained with improved dispersion.* Sensors and Actuators B: Chemical, 2006. 115(2): p. 678-684.

[36] Zhang, H., et al., Regenerated-Cellulose/Multiwalled-Carbon-Nanotube Composite Fibers with Enhanced Mechanical Properties Prepared with the Ionic Liquid 1-Allyl-3-methylimidazolium Chloride. Adv. Mater. , 2007. 19: p. 698–704.

[37] Deitzel, J.M., et al., The effect of processing variables on the morphology of electrospun nanofibers and textiles. Polymer, 2001. 42(1): p. 261-272.

[38] Zhang, S., Shim, W.S., Kim, J. Design of ultra-fine non-wovens via electrospinning of Nylon 6: Spinning parameters and filtration efficiency. Materials & Design, 2009. 30(9): p. 3659-3666.

[39] Li, Y., Huang, Z., Yandong, *Electrospinning of nylon-6,66,1010 terpolymer.* European Polymer Journal, 2006. 42(7): p. 1696-1704.

In: Development of Nanotechnology in Textiles ISBN: 978-1-62081-030-9
Editors: A. K. Haghi and G. E. Zaikov ©2012 Nova Science Publishers, Inc.

Chapter 4

COMBUSTION AND THERMAL DEGRADATION OF POLYPROPYLENE IN THE PRESENCE OF MULTIWALLED CARBON NANOTUBE COMPOSITES

G. E. Zaikov[1], S. M. Lomakin[1], N. G. Shilkina[2] and R. Kozlowski[3]

[1] NM Emanuel Institute of Biochemical Physics of Russian Academy of Sciences, Moscow, Russia
[2] NN Semenov Institute of Chemical Physics of Russian Academy of Sciences, Moscow, Russia
[3] Institut Inzynierii Materialow Polimerowych I Barwnikow, Torun, Poland

INTRODUCTION

At the present time, great attention is given to the study of properties of polymeric nanocomposites produced on the basis of well-known thermoplastics (PP, PE, PS, PMMA, polycarbonates, polyamides) and carbon nanotubes (CN). CNs are considered to have the wide set of important properties like thermal stability, reduced combustibility, electroconductivity, etc. [1-7]. Thermoplastic polymer nanocomposites are generally produced with the use of melting technique [1-12].

Development of synthetic methods and the thermal characteristics study of PP/multiwalled carbon nanotube (MWCNT) nanocomposites were taken as an objective in this paper.

A number of papers pointed at synthesis and research of thermal properties of nanocomposites (atactic polypropylene (aPP)/MWCNT) were reported [10-12]. It is remarkable that PP/MWCNT composites with minor levels of nanocarbon content (1-5% by weight) were determined to obtain an increase in thermal and thermal-oxidative stability in the majority of these publications.

Thermal stability of aPP and aPP/MWCNT nanocomposites with the various concentrations of MWCNT was studied in the paper [10]. It was shown that thermal degradation processes are similar for aPP and aPP/MWCNT nanocomposites and initial degradation temperatures are the same. However, the maximum mass loss rate temperature of PP/MWCNT nanocomposites with 1 and 5% wt of MWCNT raised by 40° - 70°C as compared with pristine aPP.

Kashiwagi et al. published the results of a study of thermal and combustion properties of PP/MWCNT nanocomposites [11, 12]. A significant decrease of a maximum heat release rate was detected during combustion research with use of a cone calorimeter. A formation of char network structure during the combustion process was considered to be the main reason of combustibility decrease. The carbonization influence upon combustibility of polymeric nanocomposites was widely presented in literature [10-12, 13]. Notably, Kashiwagi et al. [11, 12] were the first to hypothesize that abnormal dependence of maximum heat release rate upon MWCNT concentration is closely related with thermal conductivity growth of PP/MWCNT nanocomposites during high-temperature pyrolysis and combustion.

EXPERIMENTAL

Materials

Isotactic polypropylene (melting flow index = 0.7 g/10 min) was used as a polymer matrix in this paper. Multiwalled carbon nanotubes (MWCNT) (purchased from Shenzhen Nanotechnologies Co. Ltd.) were used as carbon-containing nanofillers. This product contains low amounts of amorphous carbon (less than 0.3 wt%) and could be produced with different size characteristics - different lengths and different diameters and therefore

CONE 2000 cone-calorimeter (FTT). An external radiant heat flux of 35 kW/m^2 was applied. All of the samples having a standard surface area of 70×70 mm and identical masses of 13.0±0.2g were measured in the horizontal position and wrapped with thin aluminum foil except for the irradiated sample surface.

Heat capacity and heat conductivity were determined with the use of NETZSCH 457 MicroFlash.

The electron paramagnetic resonance spectroscopy (EPR) measurements were performed in air with the PP/MWCNT (10 wt%) samples using a Mini-EPR SPIn Co. Ltd spectrometer with 100 kHz field modulation. The g factor and EPR intensity (X-band) were measured with respect to a standard calibrating sample of Mn^{2+} and ultramarine.

RESULTS AND DISCUSSION

Nanocomposite Structure

Dispersion analysis of MWNT in nanocomposites. PP/MWNT nanocomposites with original and modified MWNT were produced. Filler concentration varied from 1 to 7 wt% weight percent (0.5-3.5 volume percent, correspondingly). Distribution patterns for composites with modified and non-modified nanotubes were studied with TEM methods (Figure 2). According to TEM images, the addition of 1% by weight leads to sufficiently uniform distribution. However, agglomeration of nanotubes was detected for more concentrated nanocomposites, especially for PP/MWNT with an MWNT average diameter less than 10 nm (K1).

TEM images for PP nanocomposites with 5 wt% modified and non-modified MWNT (2.5% by volume) are shown in Figure 2. According to Figure 2, it could be stated that modified nanotubes (K2 and K3) used during the melding process are present as individual particles in nanocomposite in most cases. The number and size of agglomerates is reduced due to increased organophility and improved thermodynamic compatibility with non-polar polymer.

However, preliminary modification does not lead to uniform filler distribution for nanocomposites containing thin nanotubes (K1). This could be explained by the fact that interaction energy of CNT is more dependent on nanotube diameter than of its length. Molecular dynamic computation given in paper [24] showed that blending polymers with nanotubes becomes more

different diameter-to-length ratios. Size characteristics for three MWCNT used in this paper are given in Table 1. Sizes and structure of initial MWCNT were additionally estimated by SEM (fig.1).

Table 1. Properties of MWCNT

Designation	D, nm	L, µm	Density, g/cm^3	Specific surface area, m^2/g
MWCNT (K1)	<10	5-15	2	40 - 300
MWCNT (K2)	40-60	1-2	2	40 - 300
MWCNT (K3)	40-60	5-15	2	40 - 300

Nanocomposite Processing

Compositions were prepared by blending carbon nanotubes with melted polymer in a laboratory mixer Brabender at 190°C. TOPANOL® (1,1,3-tris (2-methyl-4-hydroxy-5-t-butylphenyl) butane) and DLTP (dilaurylthio dipropionate) were added in the amount of 0.3 and 0.5 wt% as antioxidants to prevent thermal-oxidative degradation during polymer processing.

A number of different covalent and non-covalent nanotube modifications (organofillization) were reported to be used to achieve greater structure similarity and therefore greater nanotube distribution in a polymer matrix [14-23]. In order to functionalize MWCNT, we used a preliminary ozone treatment of MWCNT followed by ammonolysis of epoxy groups on the MWCNT surface.

The selective ozonization of MWCNT was carried out with ozone-oxygen mixture (ozone concentration was 2.3×10^{-4} mol/L) in a bubble reactor. Then, the ammonolysis of oxidized MWCNT has been carried out by *tert*-butylamine in the ultrasonic bath (35 kHz) at 50°C for 120 min with following evaporation of *tert*-butylamine excess.

IR transmission spectra of tablet specimens of MWCNTs in KBr matrix was analyzed by using Perkin-Elmer 1725X FTIR spectrometer and the presence of the alkylamine groups at the MWCNT surface was confirmed by the appearance of the characteristic band ~1210 cm^{-1} corresponding to the valency vibration of the bond –C–N←.

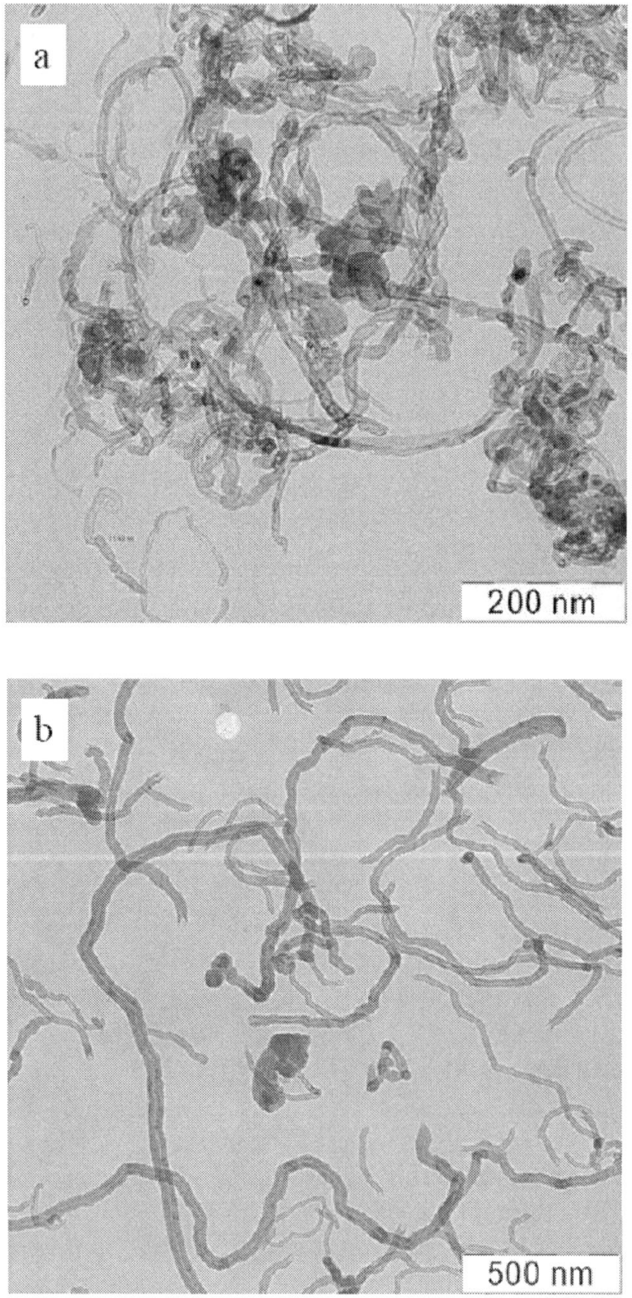

Figure 1. (Continued).

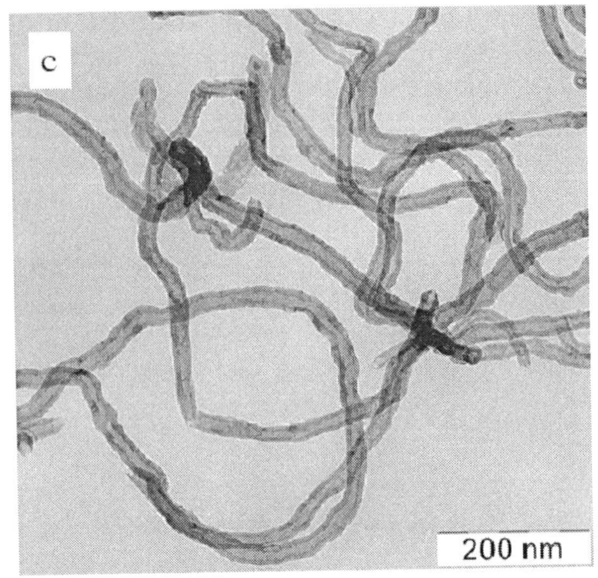

Figure 1. SEM images of original MWCNTs: (a) MWCNT(1); (b) MWCNT(2); (c) MWCNT(3).

Investigation Techniques

Scanning electron microscopy (SEM). The degree of MWCNT distribution in polymer matrix was analyzed with scanning electron microscope JSM-35. Low-temperature chips derived from film-type samples were used for this analysis.

Transmission electron microscopy (TEM). The degree of nanotube dispersion in polymer matrix was studied with transmission electron microscopy (LEO912 AB OMEGA, Germany). Microscopic sections with 70 – 100 nm width prepared with ultramicrotome "Reichert – Jung Ultracut" with a diamond cutter at -80°C. Microscopic analysis was made with accelerating potential of about 100 kV without chemical sample staining.

Thermogravimetric analysis (TG). A NETZSCH TG 209 F1 Iri thermomicrobalance has been employed for TGA measurements in oxidizin (oxygen) atmosphere. The measurements were carried out at a heating rate c 20 K/min.

Combustibility characteristics (cone calorimeter) were performe according to the standard procedures ASME E1354/ISO 5660 using a DUA

thermodynamically favorable with an increase of nanotubes diameter, owing to the fact that cohesion energy is decreased between the nanotubes and remains almost the same between the nanotubes and polymer.

Thus, mixing the thinnest nanotubes (K1) with PP leads to inevitable nanotube agglomeration in nanocomposite sample volume. Nanotube surface modification used in this paper did not result in complete overcoming of nanotube agglomeration tendency for K1 nanotubes [25]. Therefore, the PP/MWNT(K3) nanocomposites presented are the main subjects of inquiry in the present study.

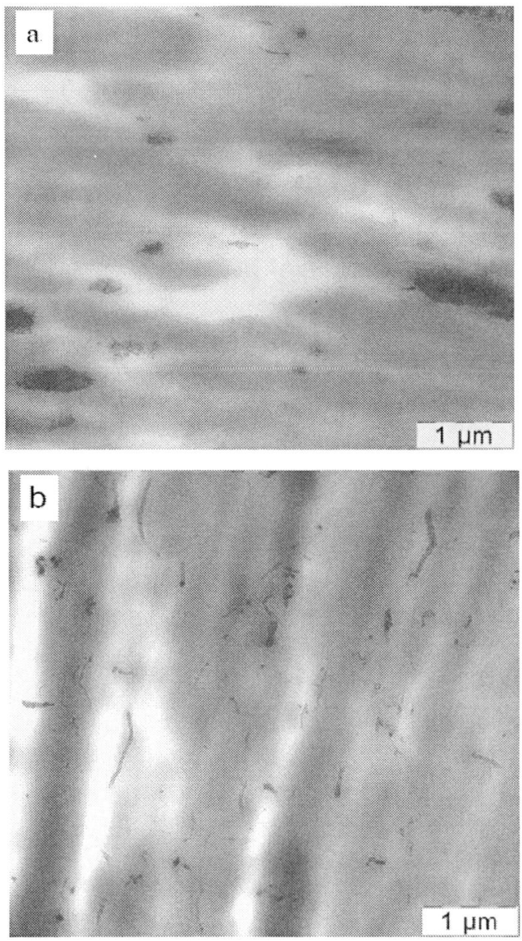

Figure 2. (Continued).

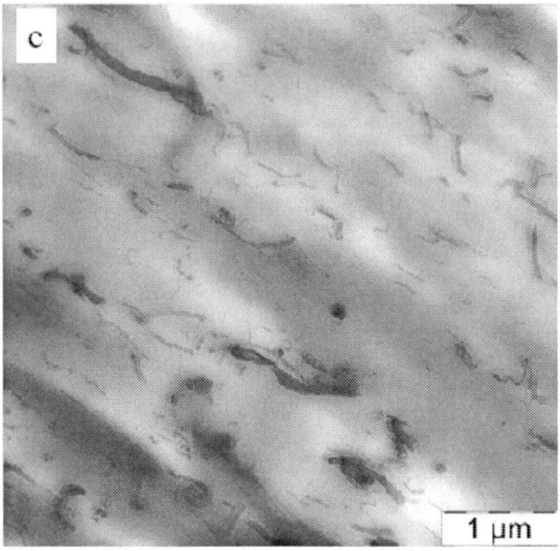

Figure 2. TEM images of PP/MWCNT nanocomposites showing dispersion of MWCNT in a polymer matrix: (a) PP/MWCNT(1); (b) PP/MWCNT(2); (c) PP/MWCNT(3).

Thermal-Oxidative Degradation of PP/MWCNT Nanocomposites

The diverse behavior of PP and PP/MWCNT nanocomposites with 1, 3 and 5 wt% of MWCNT(3). (Figure 3) shows that the influence of MWCNTs on the thermal-oxidation process resulted in higher thermal-oxidative stability of PP/MWCNT nanocomposites. It is possible to see a regular increase in the temperature values of the maximum mass loss rates (up to 60°C) for the PP/MWCNT as compared to pristine PP (Fig. 3).

Detailed analysis of TGA graphs (Figure 3) allows claiming that thermal stability increase is achieved even by an addition of 1 wt% of MWCNT to PP, while further addition does not lead to such fundamental growth. In addition, Figure 4 shows the comparative results for onset degradation temperatures ($T_{on.}$) and the maximum mass loss temperatures (T_{max}) of PP/MWCNT nanocomposites with the different types and concentrations of MWCNT. One can see non-linear relation of ($T_{on.}$) and (T_{max}) vs. MWCNT concentration in the PP compositions (Figure 4).

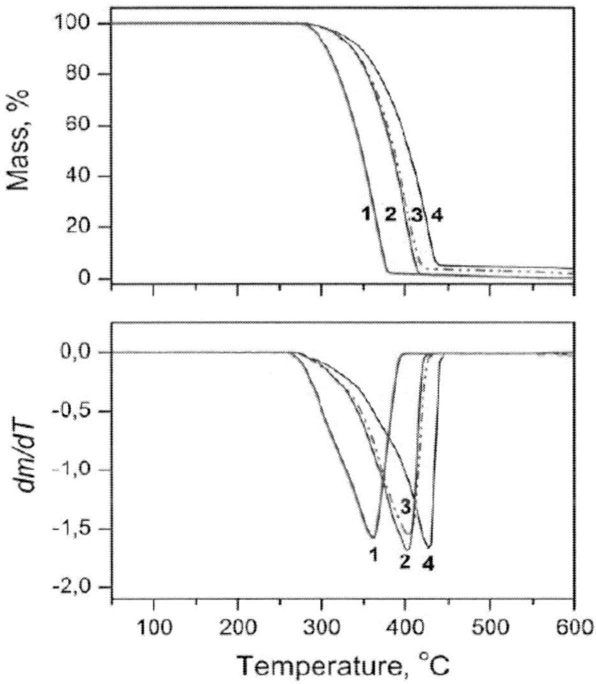

Figure 3. TG and DTG curves for PP (1) and PP/MWCNT(3) composites with 1 (2), 3 (3) and 5 wt% (4) filler loadings.

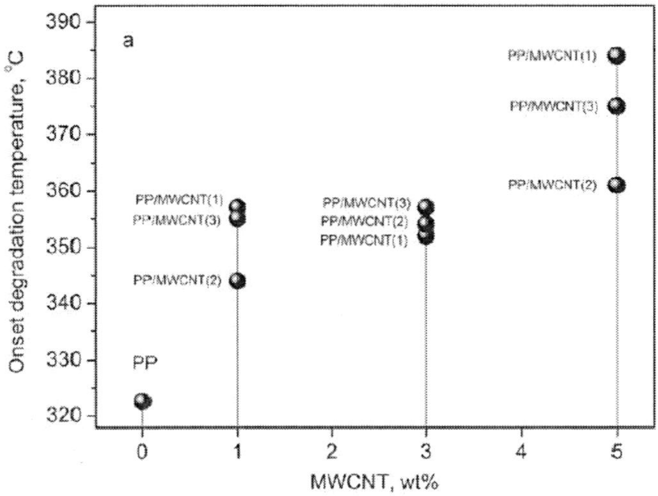

Figure 4. (Continued).

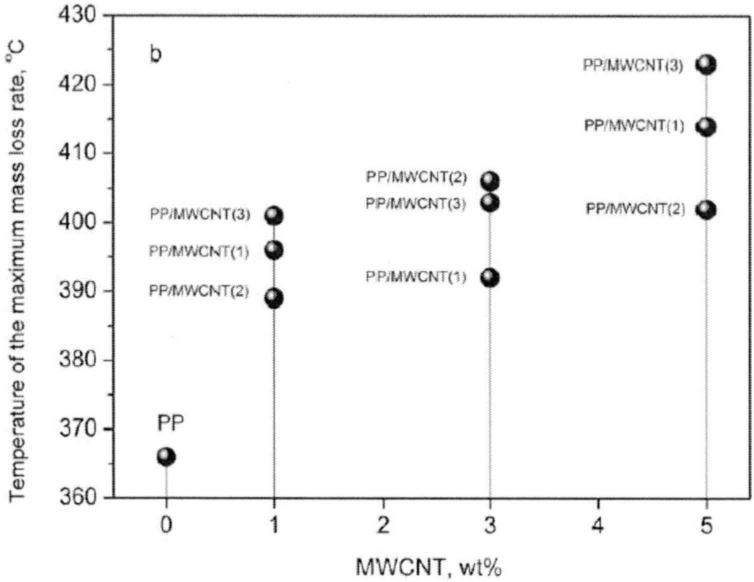

Figure 4. Comparative diagrams showing the onset degradation temperatures (a) and the maximum mass loss temperatures (b) for PP and PP/MWCNT nanocomposites with the different types and concentrations of MWCNT.

At the present time, the nature of thermal stability effect caused by MWCNT's addition to polymers is an object of comprehensive study. Most likely, MWCNTs could be considered as high temperature stabilizers (antioxidants) in reactions of thermal-oxidative degradation by analogy with fullerenes [26]. Stabilizing effect caused by an addition of MWCNT was previously detected for PP/MWCNT nanocomposites [8]: the temperature of the maximum mass loss rates of PP/MWCNT (9 wt%) was increased by 50°C as compared with pristine PP.

Results achieved in this study confirm the previous findings of the inhibiting effect of MWNCT upon the PP/MWCNT nanocomposite thermal-oxidative degradation. Obviously, the complex nature of this effect is closely related to radical-acceptor properties of MWCNT resulting in chain termination reactions of alkyl/alkoxyl radical, which leads to cross-linking and carbonization process in PP/MWCNT nanocomposites. Carbonization phenomenon was reported previously in papers aimed at PP/MWCNT heat resistance and flame retardancy study [11, 12].

Kinetic Analysis of Thermal Degradation of PP/MWNT

Kinetic studies of material degradation have a long history, and there exists a long list of data analysis techniques employed for the purpose. Often, TGA is the method of choice for acquiring experimental data for subsequent kinetic calculations, and namely this technique was employed here. It is commonly accepted that the degradation of materials follows the base equation (1) [27]

$$dc/dt = - F(t, T, c, p) \tag{1}$$

where: t - time, T - temperature, c_o - initial concentration of the reactant, and p - concentration of the final product. The right-hand part of the equation $F(t, T, c, p)$ can be represented by the two separable functions, $k(T)$ and $f(c, p)$:

$$F(t, T, c, p) = k\,[T(t)\cdot f(c, p)] \tag{2}$$

Arrhenius equation (4) will be assumed to be valid for the following:

$$k(T) = A\cdot exp(-E/RT) \tag{3}$$

Therefore,

$$dc/dt= - A\cdot exp(-E/RT)\cdot f(c,p) \tag{4}$$

All feasible reactions can be subdivided onto classic homogeneous reactions and typical solid-state reactions, which are listed in Table 2. The analytical output must provide a good fit to measurements with different temperature profiles by means of a common kinetic model.

Thermogravimetric analysis of PP and PP nanocomposite degradation was carried out in dynamic conditions at the rates of 2.5, 5 and 10 K/min on air.

Model-independent estimations of activation energy using the Friedman approach [28] was taken to get preliminary model analysis for thermal degradation and selection of initial conditions. According to this evaluation, a two-step process ($A \rightarrow X_1 \rightarrow B \rightarrow X_2 \rightarrow C$) was chosen for PP degradation. Taking into account the carbonization stage, the more complex three-step process ($A \rightarrow X_1 \rightarrow B \rightarrow X_2 \rightarrow C \rightarrow X_3 \rightarrow D$) was selected for PP/MWNT degradation [27, 29].

According to the results of non-linear regression and taking the set of reaction models into consideration, we computed the values of active kinetic parameters, which represent the best approximation of experimental TGA graphs (Figure 5, Table 3).

Table 2. Considered reaction models $dc/dt=-A\cdot exp(-E/RT)f(c, p)$

Reaction models	$f(c, p)$
First order (F_1)	c
Second order (F_2)	c^2
n-order (F_n)	c^n
Two-dimensional phase boundary (R_2)	$2\cdot c^{1/2}$
Three-dimensional phase boundary (R_3)	$3\cdot c^{2/3}$
One-dimensional diffusion (D_1)	$0.5/(1 - c)$
Two-dimensional diffusion (D_2)	$-1/ln(c)$
Three-dimensional diffusion, Jander's type (D_3)	$1.5c^{1/3}(c^{-1/3} - 1)$
Three-dimensional diffusion, Ginstling-Brounstein (D_4)	$1.5/(c^{-1/3} - 1)$
One-dimensional diffusion (Fick law) (D_{1F})	-
Three-dimensional diffusion (Fick law) (D_{3F})	-
Prout-Tompkins equation (B_1)	$c\cdot p$
Expanded Prout-Tompkins equation (B_{na})	$c^n\cdot p^a$
First order reaction with autocatalysis by X (C_{1-X})	$c\cdot (1+K_{cat} X)$
n-order reaction with autocatalysis by X (C_{n-X})	$c^n\cdot(1+K_{cat} X)$
Two-dimensional nucleation, Avrami-Erofeev equation (A_2)	$2\cdot c\cdot(-ln(c))^{1/2}$
Three-dimensional nucleation, Avrami-Erofeev equation (A_3)	$3\cdot c\cdot(-ln(c))^{2/3}$
n- dimensional nucleation, Avrami-Erofeev equation (A_n)	$n\cdot c\cdot (-ln(c))^{(n-1)/n}$

Table 3. Kinetic parameters for thermal degradation of (a) PP (Fn □ Fn) and (b) PP/MWNT nanocomposite (Fn □D1□ Fn). TGA analysis was performed in air flow with the use of multiple non-linear regression analysis for model processes

(a)

Reaction model	Kinetic parameters	Values	Correlation coefficient
Fn →Fn	$logA_1$, s^{-1}	9.53	0.9996
	E_1, kJ/mol	110.25	
	n_1	1.89	
	$logA_2$, s^{-1}	15.25	
	E_2, kJ/mol	150.65	
	n_2	1.50	

(b)

Reaction model	Kinetic parameters	Values	Correlation coefficient
Fn →D1→ Fn	$\log A_1$, s^{-1}	6.3	0.9996
	E_1, kJ/mol	105.1	
	n_1	0.91	
	$\log A_2$, s^{-1}	7.4	
	E_2, kJ/mol	120.4	
	$\log A_3$, s^{-1}	16.7	
	E_3, kJ/mol	229.5	
	n_3	0.5	

Two-step PP thermal-oxidative degradation in dynamic heating conditions was confirmed by obtained data [30]. At the first stage, the values of activation energy and pre-exponential factor are 110.25 kJ/mol and $10^{9.5}$ $s^{-1,}$ correspondingly, while the reaction order is close to 2 (1.89). The values of activation energy and pre-exponential factor are larger on the second stage (E_2 = 150.65 kJ/mol, $A_2 = 10^{15.3}$ s^{-1}) with effective reaction order of $n_2 = 1.50$.

The preferred model for PP/MWNT thermal-oxidative degradation and with respect to statistical analysis of kinetic parameters is composed of 3 consecutive reactions $F_n \rightarrow D_1 \rightarrow F_n$, where D_1 – one-dimensional diffusion, and F_n – n-order reaction (Figure 4b, Table 3b). In this case, the first step activation energy is equal to 105.1 kJ/mol, reaction order is close to 1 (n_1 = 0.91). On the second step, which is described as one-dimensional diffusion, the value of activation energy is equal to 120.4 kJ/mol, while the value is almost twice as large for the third step (E_3 =229.5 kJ/mol) with effective reaction order of $n_3 = 0.5$ (table 3b).

Comparison of thermal oxidative degradation parameters for PP/MWNT with layered silicate PP//MMT showed that the values of activation energy of the second and the third stages are higher for PP/MWNT:

E_2=120.4 kJ/mol and E_3=229.5 kJ/mol for PP/MWNT;

E_2=100.0 kJ/mol and E_3=199.8 kJ/mol for PP/MPP/MMT correspondingly [30].

This data may testify to more intensive carbonization in cases of PP/MWNT than in case of PP/MMT, which finally leads to a decrease in RHR value.

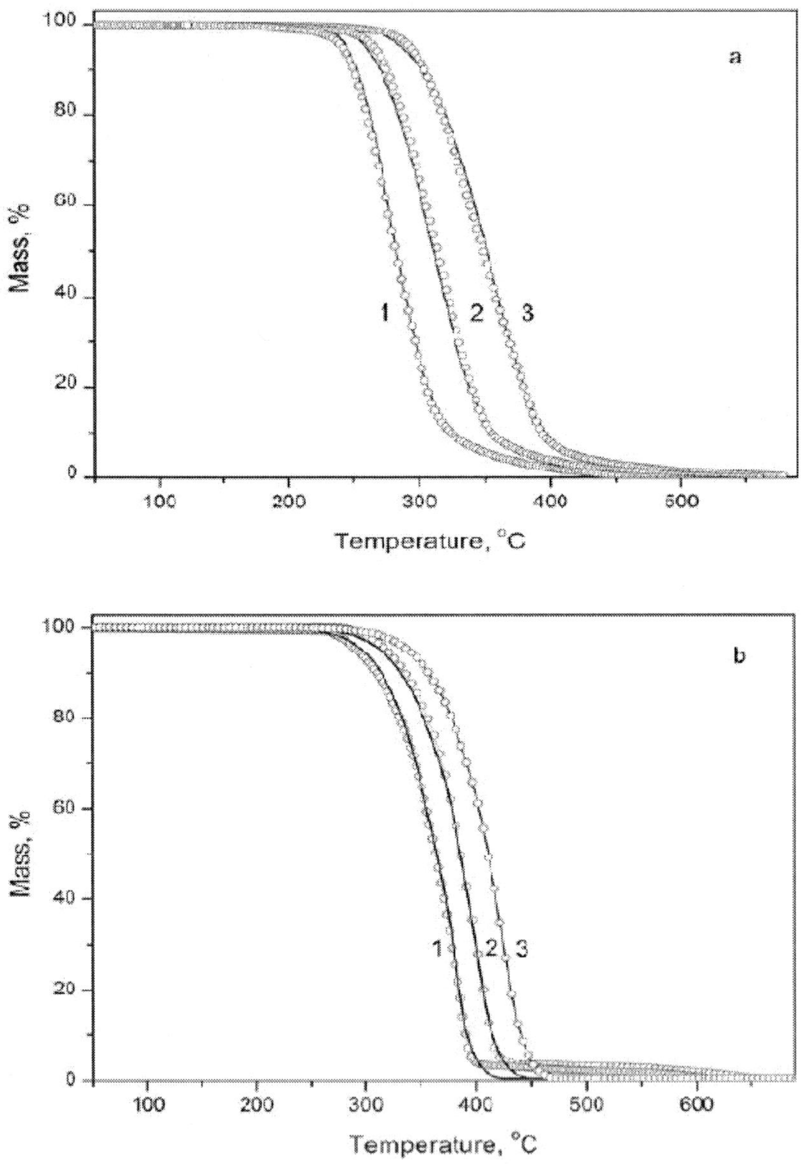

Figure 5. Non-linear kinetic modeling of a – PP and b – PP/MWCNT(3) thermal-oxidative degradation in air. Comparison between experimental TG data (dots) and the model results (firm lines) at several heating rates: (1) – 2.5, (2) – 5, (3) – 10 K/min.

COMBUSTIBILITY OF PP/MWCNT NANOCOMPOSITES

Figure 6 depicts the plots of the heat release rate (RHR), as basic flammability characteristic, vs. time for PP, as well as for the PP/MWCNT nanocomposites.

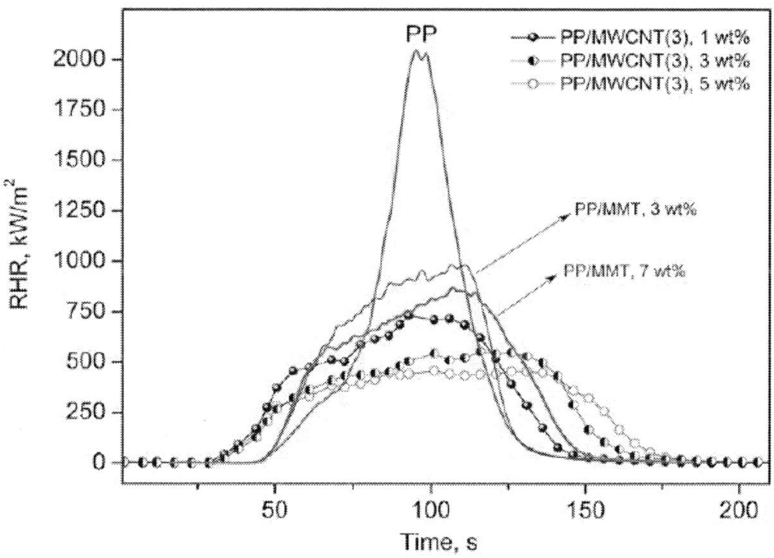

Figure 6. Rate of heat release vs. time for PP, PP/MWCNT and PP/MMT(Cloisite 20A) nanocomposites obtained by cone calorimeter at the incident heat flux of 35 kW m^{-2}.

From Figure 6, it could be seen that the maximum heat release rate for pristine PP is 2076 kW/m^2, whereas that for the PP/MWCNT %) nanocomposites (1 wt%), PP/MCWNT (3 wt%) and the PP/MCWNT (5 wt.%) RHR values are 729 kW/m^2, 552.8 kW/m^2, and 455.8 kW/m^2, respectively; thus, the peak heat release rate decreases by 65%, 73% and 78%.

The observed flame retardancy effect is associated with solid-phase carbonization reactions, by analogy with layered silicates [31, 32]. In an early paper [30], we have found that additions of 3 and 7 wt% of layered silicate (Cloisite 20A) to the PP compositions PP lead to RHR decrease by 51 and 57% as compared with pristine PP (Figure 5).

We believe that a higher carbonization effectiveness of MWCNTs depends on their heat conductivity. It is well known that PP has a low thermal conductivity at standard conditions, and characterized by a minor increase with temperature up to melting point (~0.2 W/m K). On the other hand, the heat conductivity of individual MWCNT is extremely high and equals to 3000 W/m K [33, 34]. During the high temperature pyrolysis of PP/MWCNT composition at temperatures above 300-400°C, the heat conductivity can rise up to 20W/m K [34] due to an actual increase of MWCNT concentration in composition caused by volatilization of polypropylene degradation products (Figure7). The induced heating of PP/MWCNT intensifies a steady carbonization and charring of the samples and leads to decrease of RHR peak value (Figure 6).

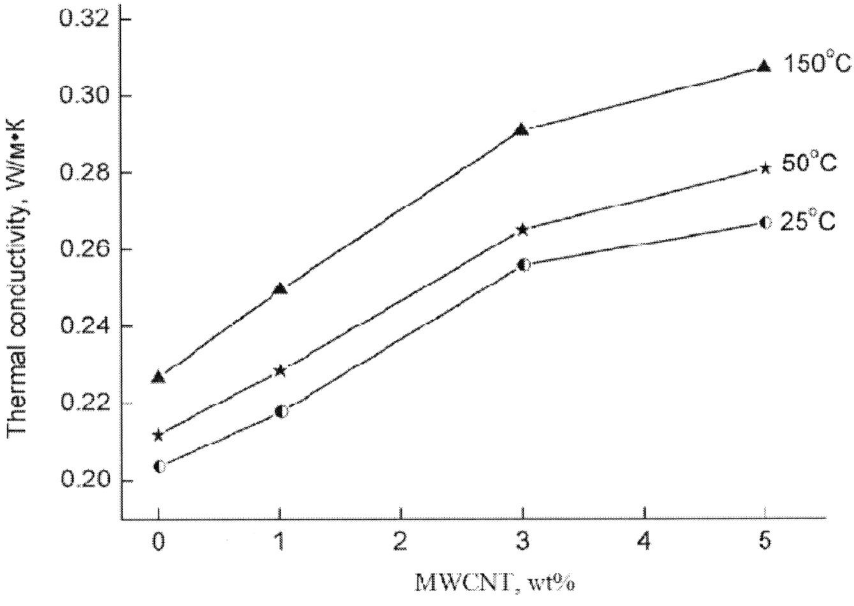

Figure 7. Temperature dependence of the thermal conductivity of PP/MWCNT (3) nanocomposite with different loadings of MWCNT.

Figures 8 and 9 show graphs for the specific extinction area and effective heat of combustion, correspondingly, for PP and PP/MWCNT(3) nanocomposites. Calculated values of effective heat of combustion for PP and PP/MWNT demonstrate an invariant shift of this parameter for these nanocomposites.

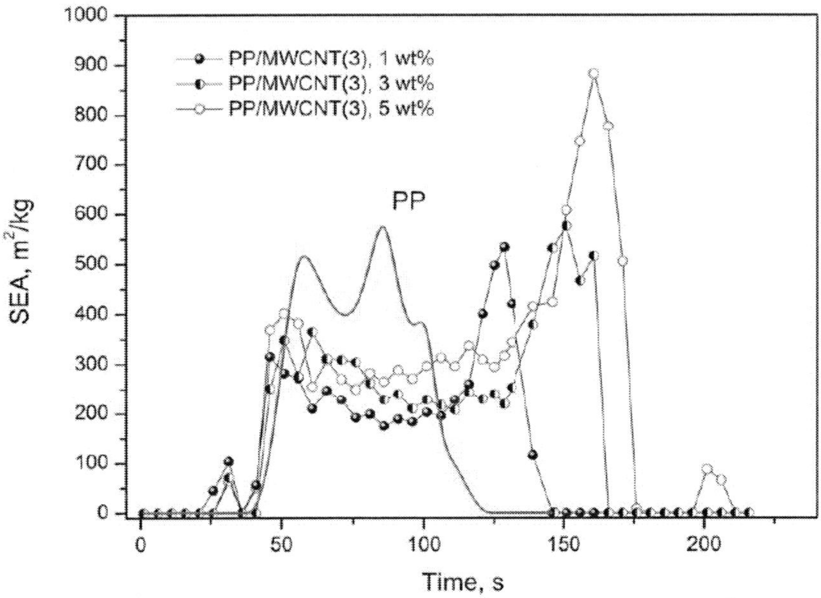

Figure 8. Specific extinction area vs. time for PP and PP/MWCNT(3) nanocomposites obtained by cone calorimeter at the incident heat flux of 35 kW m^{-2}.

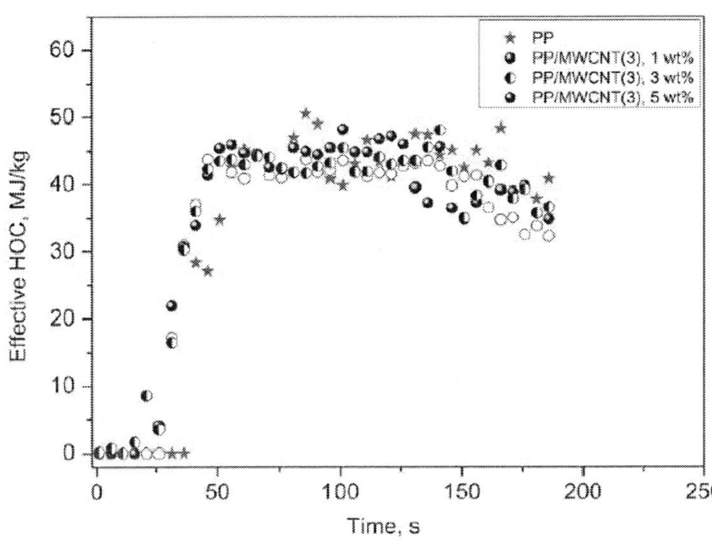

Figure 9. Effective heat of combustion vs. time for PP and PP/MWCNT(3) nanocomposites obtained by cone calorimeter at the incident heat flux of 35 kW m^{-2}.

In the present study, EPR research was performed to follow the formation of stable radicals, responsible for the carbonization process, upon isothermal heating of PP/MCWNT (10 wt%) in air at 350°C.

Figure 10a shows EPR spectrum of the stable paramagnetic centers formed in the samples of PP/MCWNT (10% wt) heating in air at 350°C. When heated in air, the PP/MCWNT specimen was placed into an EPR sample tube, a narrow singlet signal with a line width of $\Delta H_{1/2}$ = 0,69 mT and a g value of 2.003 was detected due to the stable radicals generation, analogous to those previously registered during polymers carbonization process [35]. No EPR signal similar to that of PP/MCWNT samples were observed in the samples of pristine PP and MCWNT samples heated at 350°C in air. It should be noted that although iron impurity from MWNCT has been mentioned in other studies on pyrolysis of polymer nanocomposites as the radical traps [36, 11], the EPR analyses in the current study showed the presence of paramagnetic centers relating to carbonaceous stable radicals only.

As it is seen from Figure 10b, the formation of stabilized radicals occurs with a pronounced induction period which is related to antioxidant properties of MWNCT. Such a type of kinetic dependence is coincided with an oxygen uptake kinetics observed during inhibited polyolefines thermal oxidation. Moreover, no EPR signals were observed in the samples of the PP/MCWNT samples heated at 350°C in inert Ar.

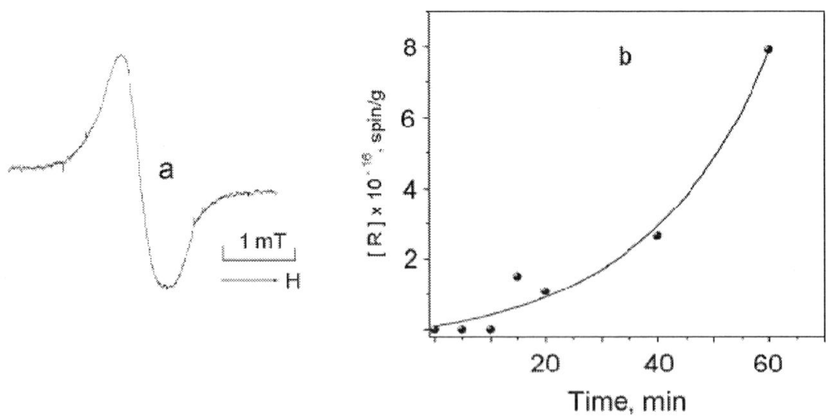

Figure 10. EPR spectrum of the stable paramagnetic centers formed in the samples of PP/MCWNT (10% wt) heating in air at 350°C – (a), kinetic dependence of stable radicals generation from PP/MCWNT (10% wt) under isothermal heating at 350°C in air – (b).

Thus, multiwalled carbon nanotubes are considered to be more effective filling agents than layered silicates in the terms of improvement of thermal properties and flame retardancy of PP matrix. This could be explained by the specific antioxidant properties and high thermal conductivity of MWCNT, which determine the carbonization reactions during the thermal-oxidative degradation process.

REFERENCES

[1] Shaffer, M.S.P., Windle, A.H. *Adv. Mater.* 1999, 11, 937.

[2] Qian, D., Dickey, E.C., Andrews, R., Rantell, T. *Appl. Phys. Lett.* 2000, 76, 2868.

[3] Jin, Z., Pramoda, K.P., Xu, G., Goh, S.H. *Chem. Phys. Lett.* 2001, 337, 43.

[4] Thostenson, E.T., Chou, T.W. *J. Phys. D: Appl. Phys*. 2002, 35, L77.

[5] Bin, Y., Kitanaka, M., Zhu, D., Matsuo, M. Macromolecules 2003, 36, 6213.

[6] Potschke, P., Dudkin, S.M., Alig, I. *Polymer.* 2003, 44, 5023.

[7] Safadi, B., Andrews, R., Grulke, E.A. *J. Appl. Polym. Sci.* 2002, 84, 2660.

[8] Watts, P.C.P., Fearon, P.K., Hsu, W.K., Billingham, N.C., Kroto, H.W., Walton, D.R.M. *J. Mater. Chem*. 2003, 13, 491

[9] Watts, P.C.P., Hsu, W.K., Randall, D.P., Kroto, H.W.,Walton, D.R.M. *Phys. Chem. Chem. Phys.,* 2002, 4, 5655.

[10] Jun Yang, Yuhan Lin, Jinfeng Wang. *J. Appl. Polym. Sci.* 2005, 98,3,1087

[11] Kashiwagi, T., Grulke, E., Hilding, J., Groth, K., Harris, R.H., Butler, K., Shields, J.R., Kharchenko, S., Douglas, J. *Polymer.* 2004, 45, 4227–4239

[12] Kashiwagi, T., Grulke, E., Hilding, J., Harris, Jr. R.H., Awad, W.H., Douglas, J. *Macromol. Rapid Commun.* 2002, 23, 761, 5.

[13] Lomakin, S.M., Novokshonova, L.A., Brevnov, P.N., Shchegolikhin, A.N. *Journal of Materials Science*. 2008, 43, 4, 1340.

[14] Chen, J., Hamon, M.A., Hu, H., Chen, Y., Rao, A.M., Eklund, P.C., Haddon, R.C. *Science.* 1998, 282, 95.

[15] Stevens, J.L., Huang, A.Y., Peng, H., Chiang, I.W., Khabashesku, V.N., Margrave, J.L. *Nano Lett.,* 2003, 3, 331.

[16] Eitan, A., Jiang, K., Dukes, D., Andrews, R.. Schadler, L.S. *Chem. Mater.,* 2003, 15 3198.
[17] Hu, H., Ni, Y., Montana, V., Haddon, R.C., Parpura, V. *Nano Lett.,* 2004, 4, 507.
[18] Kong, H., Gao, C., Yan, D.J. *Am. Chem. Soc.,* 2004,126, 412.
[19] Holzinger, M., Vostrowsky, O., Hirsch, A., Hennrich, F., Kappes, M., Weiss, R., Jellen, F. Angew. *Chem. Int.* Edn 2001,40, 4002.
[20] Holzinger, M., Abraham, J., Whelan, P., Graupner, R., Ley, L., Hennrich, F., Kappes, M., Hirsch, A. J. *Am. Chem. Soc.,* 2003, 125, 8566.
[21] Yao, Z., Braidy, N., Botton, G.A., Adronov, A. 2003 *J. Am.Chem. Soc.* 125, 16015.
[22] Ying, Y., Saini, R.K., Liang, F., Sadana, A.K., Billups, W.E. *Org. Lett.* 2003, 5, 1471.
[23] Alvaro, M., Atienzar, P., de la Cruz, P., Delgado, J.L., Garcia, H., Langa, F. 2004 *J. Phys. Chem.* B 108 12691,
[24] Nyden, M.R., Stoliarov, S.I. *Polymer.* 2008,49, 635-641,
[25] Rakhimkulov, A.D., Lomakin, S.M., Alexeyeva, O.V., Dubnikova, I.L., Schegolikhin, A.N., Zaikov, G.E. in *Proceedings of IBCP International Conference* (2006) 56.
[26] Krusic, J., Wasserman, E., Keizer, P.N., Morton, J.R., Preston, K.F. *Science.* 1991, 254, 1183,
[27] Opfermann, J. *J. Thermal Anal Cal.* 2000, 60, 3, 641.
[28] Friedman, H.L. *J. Polym. Sci.,* 1965, C6, 175.
[29] Opfermann, J., Kaisersberger, E. *Thermochim. Acta.* 1992, 11, 1, 167.
[30] Lomakin, S.M., Dubnikova, I.L., Berezina, S.M., Zaikov, G.E. *Polymer Science.* 2006, 48, Ser. A, 1, 72.
[31] Gilman, G.W., Jackson, C.L., Morgan, A.B., Harris, R.H., Manias, E., Giannelis, E.P., Wuthenow, M., Hilton, D., Phillips, S. *Chem. Mater.* 2000, 12, 1866.
[32] Kashiwagi, T., Harris, R.H. Jr., Zhang, X., Briber, R.M., Cipriano, B.H., Raghavan, S.R., Awad, W.H., Shields, J.R. *Polymer.* 2004, 45, 881.
[33] Kim, P., Shi, L., Majumdar, A., McEuen, P.L. *Phys. Rev. Lett.* 2001, 87, 215502.
[34] Yi, W., Lu, L., Zhang, D.L., Pan, Z.W., Xie, S.S. *Phys. Rev. B* 1999, 59, R9015–8.
[35] Echevskii, G.V., Kalinina, N.G., Anufrienko, V.F., Poluboyarov, V.A. *React. Kinet. Catal. Lett.* 1987, 33, 8, 305.

[36] Zhu, J., Uhl, F., Morgan, A.B., Wilkie, C.A. *Chem. Mater.* 2001, 13,4649–54.

In: Development of Nanotechnology in Textiles ISBN: 978-1-62081-030-9
Editors: A. K. Haghi and G. E. Zaikov ©2012 Nova Science Publishers, Inc.

Chapter 5

POLYMER/ORGANOCLAY NANOCOMPOSITES REINFORCEMENT

***Georgiy V. Kozlov*[1], *Boris Zh. Dzhangurazov*[1], *Stefan Kubica*[2], *Gennady E. Zaikov*[3] *and Abdulakh K. Mikitaev*[1)]**

[1] Kabardino-Balkarian State University, Russia
[2] Institut Inzynierii Materialow Polimerowych I Barwnikow, Torun, Poland
[3] N.M. Emanuel Institute of Biochemical Physics of Russian Academy of Sciences, Russia

INTRODUCTION

At present, nanocomposites polymer/organoclay studies attained very big widespreading [1, 2]. However, the majority of works fulfilled on this theme has mainly an applied character and theoretical aspects of polymers reinforcement by organoclays are studied much less. So, the authors [3] developed for this purpose multiscale micromechanical model, in the basis of which representation about organoclay "effective particle" was assumed. The indicated "effective particle" includes in itself both nanofiller platelets and adjoining to them (or located between them) polymeric matrix layers. Despite the complexity of model [3], it has an essential lacks number. First, the indicated above complexity results to the necessity of parameters large number usage, a part of which is difficult enough and sometimes impossible to determine. Secondly, this model is based on a micromechanical models application, which in essence exhausted their resources [4]. Thirdly, at the entire of its complexity, the model [3] takes not into account such basis for polymer nanocomposites factors as a real level of interfacial adhesion nanofiller-polymeric matrix and polymer chain flexibility for nanocomposite matrix.

The other model [2], used for the same purpose, is based on the following percolation relationship application [5]:

$$\frac{E_n}{E_m} = 1 + 11\varphi_n^{1.7}, \tag{1}$$

where E_n and E_m are elasticity modulus of nanocomposite and matrix polymer, accordingly, φ_n is nanofiller volume contents.

Let us note, that the model [2] is based on the approach, principally different from micromechanical models: it is assumed that polymer composites properties are defined by their matrix structural state only and the filler role consists in modification and fixation of matrix polymer structure [6].

Proceeding from the said above, the model [2] correctness checking is the present purpose of the paper with using the experimental data for nanocomposites epoxy polymer/Na^+-montmorillonite (EP/MMT) [7] and also the comparison of models [2] and [3] theoretical estimations for the mentioned nanocomposites.

EXPERIMENTAL

A matrix polymer epoxy polymer on the basis of 3,4-epoxycyclohexylmethyl-3,4-epoxycyclohexane carboxylate (epoxy monomer), cured by hexahydro-4-methylphalic anhydride at molar ratio 0.87-1.0, was used. To this mixture, nanofiller Na^+-montmorillonite, under product name Cloisite 30B, was added. Beforehand, the nanofiller was mixed with denatured ethanol for the removal of any excess of surfactants from silicate platelets surface. This mixture was then centrifuged at 5000 rpm for 10 min prior to decanting of the ethanol. After the Cloisite 30B addition, the mixture was stirred in a speedmixer DAC 150 FV at a setting of 2500 rpm for 45 s and then at 3000 rpm for the same amount of time [7].

The nanocomposites EP/MMT samples were prepared in three stages. The samples were first cured isothermally for up to 8 h at temperatures 353, 373,

393 and 413 K, followed by 8 h at 453 K, and finally, 12 h at 493 K under vacuum. The samples were cooled down at a rate of 1-2 K/min and 20×5×1 mm^3 pieces for dynamic spectroscopy measurements were machined. Na^+-montmorillonite contents in the studied nanocomposites made up of 2, 5, 10 and 15 mass% [7].

The dynamical mechanical tests were carried out by using a Rheometric Scientific ARES rheometer in torsional mode at a frequency of 1 s^{-1} in nitrogen atmosphere.

The nanocomposites structure studies using wide-angle X-ray diffractometry were carried out on Bruker-AXS General Area Diffraction Detection System using Cu irradiation with wave length λ=1.54 Å. The distance between Na^+-montmorillonite platelets (d_{001} spacing) was calculated according to these measurements [7].

RESULTS AND DISCUSSION

The model [3] essence consists of the following. Composite continuous simulation found out that properties improvement depends strongly on filler particles individual features: their volume fraction f_p, particle aspect ratio (anisotropy) L/t and ratio of particle and matrix mechanical properties. These important aspects of nanocomposites polymer/organoclay require coordination and precise definition. The authors [3] used multiscale simulation strategy for the calculation of nanocomposite hierarchical morphology at a scale of order of a thousand microns. The large particle aspect ration within the matrix limits represents the structure; at a scale of several microns, the clay particle structure presents itself either fully divided organoclay platelets with a thicknes of order of a nanometer, or organoclay parallel platelets packing, separated by interlayer galleries with a thickness of nanometer level, and polymeric matrix. In this case, the quantitative structural parameters, obtained by X-ray diffraction and electron microscopy methods (silicate platelets number N in organoclay bundle, spacing d_{001} between silicate platelets) were used by the authors [3] for definition of clay "particles" geometrical features including L/t and ratio f_p to nanofiller mass content W_n. These geometrical features together with silicate platelet stiffness estimations, obtained from molecular dynamics simulations, give basis for prediction of organoclay particle effective mechanical properties. It is easy to see that it is just such approach that defines the model [3] complexity.

Unlike many mineral fillers used at plastics production (talcum powder, mica ect.), organoclays, in particular, montmorillonite, are capable to stratify and disperse into separate platelets with thickness of about 1 nm [3]. The montmorillonite platelets bundles, inseparable after the introduction into the polymer, are often called tactoids. The term "intercalation" describes the case, when small polymer amounts penetrate into galleries between silicate platelets, that causes these platelets separation on the value ~ 2-3 nm. Exfoliation or stratification occurs at the distance between platelets (in X-raying this distance is accepted to be called the spacing d_{001}) of an order of 8-10 nm. Well-stratified and dispersed nanocomposite includes separate organoclay platelets, uniformly distributed in polymeric matrix [1]. Equation (1) usage for the prediction of nanocomposites reinforcement degree E_n/E_m has shown the necessity of its following modification [1, 2]. First of all, it has been found out that not only in the full sense nanofiller (organoclay platelets), but also interfacial regions, formed on its platelets surface, with relative fraction φ_{if} are nanocomposites structure reinforcing or strengthening element. Such a situation is due to interfacial layers higher stiffness in comparison with bulk polymeric matrix in virtue of strong interactions polymer – organoclay and molecular mobility suppression in the mentioned layers [2]. Thus, Na^+-montmorillonite platelets and formed on their surface interfacial regions also presents themselves as organoclay "effective particles" with volume concept (φ_n+φ_{if}), where φ_n are nanofiller volume contents. Then equation (1) can be rewritten as follows [1]:

$$\frac{E_n}{E_m} = 1 + 11\left(\varphi_n + \varphi_{if}\right)^{1.7}. \qquad (2)$$

Besides, within the frameworks of fractal model of interfacial layers formation in polymer nanocomposites, it has been shown that between φ_{if} and φ_n, the following relationship exists [8]:

$$\varphi_{if} = 0.955\varphi_n b \qquad (3)$$

for intercalated organoclay and

$$\varphi_{if} = 1.910\varphi_n b \qquad (4)$$

for the exfoliated one, where b is parameter, characterizing the nanofiller – polymeric matrix interfacial adhesion level [9]. Let us note that equations (3) and (4) were obtained by accounting for organoclay platelets strong anisotropy. For example, for particulate (approximately spherical) nanofiller particles, the similar equation has the following form [2]:

$$\varphi_{if} = 0.102\varphi_n b\,. \tag{5}$$

The equations (2)-(4) combination allows to obtain the final variant of the formulae for nanocomposites polymer/organoclay reinforcement degree determination [2]:

$$\frac{E_n}{E_m} = 1 + 11(1.955\varphi_n b)^{1.7} \tag{6}$$

for intercalated organoclay and

$$\frac{E_n}{E_m} = 1 + 11(2.910\varphi_n b)^{1.7} \tag{7}$$

for the exfoliated one.

Therefore, the equations (2)-(7) comparison demonstrates that the model [2] accounts for nanofiller volume contents, its particles anisotropy degree and interfacial adhesion level, but does not account for nanofiller characteristics (for example, its elasticity modulus), that differs this model from the treatment [3].

The nanofiller volume fraction φ_n can be estimated as follows. At it is known [5], the following interconnection exists between nanofiller volume φ_n and mass W_n contents:

$$\varphi_n = \frac{W_n}{\rho_n}, \tag{8}$$

where ρ_n is the nanofiller density, determined according to the equation [5]:

$$\rho_n = \frac{6}{S_u D_p}, \tag{9}$$

where S_u is the filler specific surface, which is equal to ~ 74×10^3 m^2/kg for Na^+-montmorillonite [11], D_p is its particles size. Since Na^+-montmorillonite particle length is anisotropic and has length is ~ 100 nm, width is ~ 35 nm and thickness is ~ 1 nm [10], then as D_p, these sizes' arithmetical mean was chosen. Then, the value ρ_n=1790 kg/m^3. In paper [7], it has been found out experimentally that for nanocomposites EP/MMT with W_n=2 and 5 mass %, the interlayer spacing value $d_{001}\geq10$ nm and for the same nanocomposites with W_n=10 and 15 mass. % d_{001}=6.3-9.8 nm. Therefore, according to the above classification, nanocomposites with W_n=2 and 5 mass % are to be defined as having exfoliated organoclay and with W_n=10 and 15 mass % – as intercalated organoclay. In Figure 1, the curves $E_n(W_n)$, calculated for the indicated above cases according to the equations (7) and (6), accordingly, at the condition b=1 (perfect adhesion by Kerner) and also the experimental data for nanocomposites EP/MMT (points) are adduced. As one can see, the good correspondence of the data for EP/MMT with W_n=10 and 15 mass % and calculation according to the equation (6) and with W_n=2 and 5 mass. % – to the equation (7), that is provided for the model [2]. Therefore, the mentioned model reflects both nanofiller contents and its platelets stratification degree and in addition, it is much simpler and physically much clearer than the multiscale model [3]. One more aspect of the model [2] concerns the parameter b determination, which characterizes interfacial adhesion nanofiller – polymeric matrix level. The condition b=1 was accepted for the value E_n estimation (Figure 1), but this parameter can also be estimated more precisely according to the equations (6) and (7), using experimental values E_n, E_m and φ_n independent estimation of interfacial adhesion level for the studied nanocomposites can be obtained with the aid of the parameter A, determined according to equation [12]:

$$A = \frac{1}{1-\varphi_n}\frac{\mathrm{tg}\,\delta_n}{\mathrm{tg}\,\delta_m} - 1, \tag{10}$$

where tg δ_n and tg δ_m are mechanical, it loses angle tangents for nanocomposite and matrix polymer, accordingly, the values of which are adduced in paper [7].

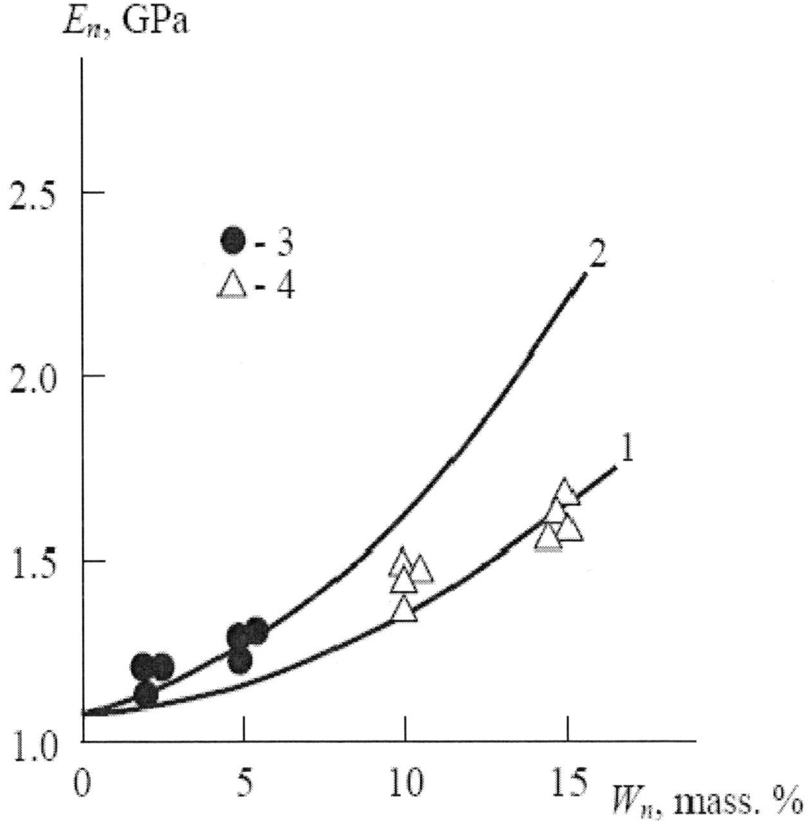

Figure 1. The comparison of calculated according to the equations (6) (1) and (7) (2) and experimental of exfoliated (3) and intercalated (4) organoclay dependences of elasticity modulus E_n on organoclay mass contents W_n for nanocomposites EP/MMT.

As it is known [12], A decrease means interfacial adhesion level increase. In Figure 2, the comparison of parameters A and b, characterizing interfacial adhesion level, is adduced for nanocomposites EP/MMT.

As it has been expected, A decrease according to the indicated above reasons, corresponds to b growth, i.e., determined according to the equations (6) and (7) parameter b characterizes the real level of interfacial adhesion in nanocomposites polymer/organoclay. As it was assumed above, the greater part obtained by the indicated mode values b is close to a unit.

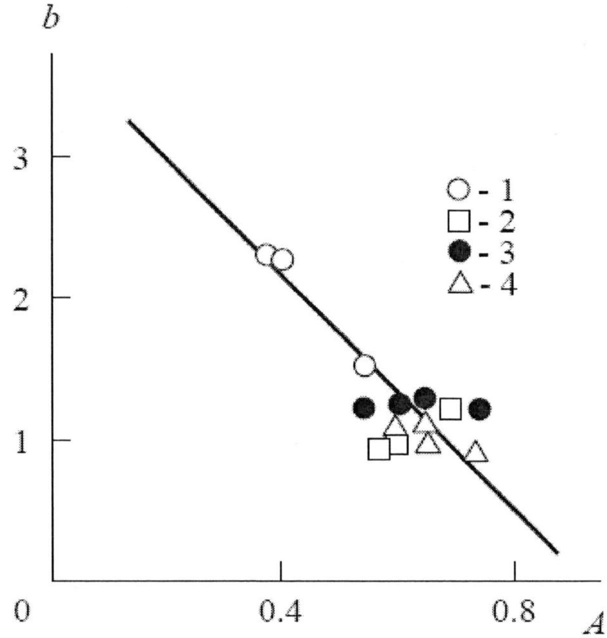

Figure 2. The relation of parameters b and A, characterizing interfacial adhesion level, for nanocomposites EP/MMT with organoclay mass contents W_n=2 (1), 5 (2), 10 (3) and 15 (4) mass %.

Furthermore, let us compare organoclay "effective particle" characteristics in models [2] and [3]. The authors [3] calculated the dependences of silicate relative fraction χ in "effective particle" and plotted its dependences on ratio d_{001}/d_{pl} for different N, where d_{pl} is silicate platelet thickness (d_{pl}=1 nm), N is these platelets number in "effective particle". In Figure 3, theoretical calculations according to the model [3] are shown by solid lines for N=1, 2 and 4, and experimental dependences of $\varphi_n/(\varphi_n+\varphi_{if})$ in the EP/MMT – by points. As one can see, the similar trends of χ and $\varphi_n/(\varphi_n+\varphi_{if})$ change as a function of d_{001}/d_{pl} are obtained, that indicates the identity of "effective particle" definition in the models [2] and [3]. The comparison of theoretical curves and experimental points in Fig. 3 shows that for nanocomposites EP/MMT the value N is somewhat smaller than 2, i.e., organoclay in these nanocomposites is stratified enough.

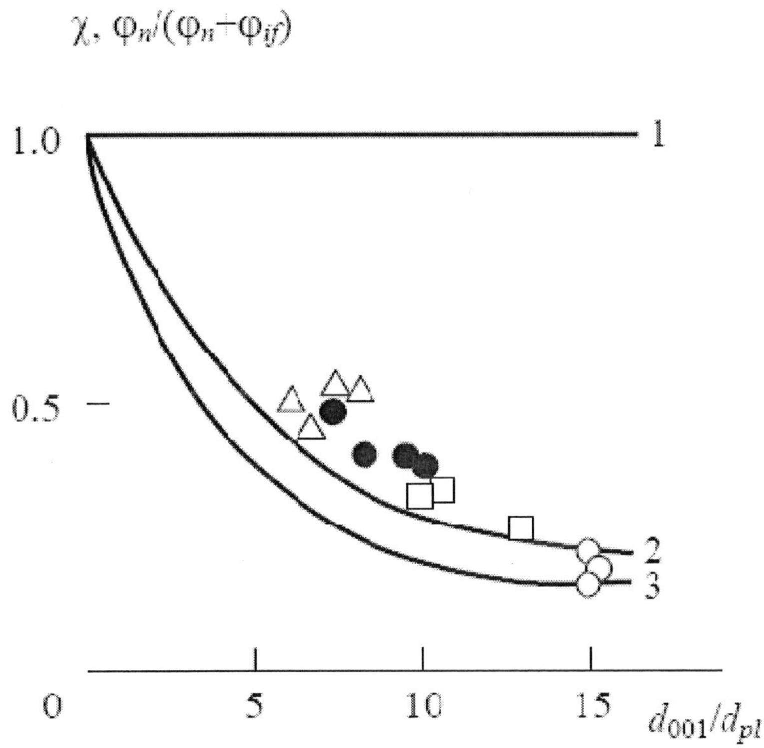

Figure 3. The theoretical dependences of organoclay fraction in "effective particle" χ for silicate platelets number in it N=1 (1), 2 (2) and 4 (3) and parameter $\varphi_n/(\varphi_n+\varphi_{if})$ for nanocomposites EP/MMT (points) on the ratio d_{001}/d_{pl} value. Conventional signs are the same as that in Figure 2.

In Figure 4, theoretical dependences of "effective porticles" volume fraction and nanofiller mass contents ratio f_p/W_n on ratio d_{001}/d_{pl} value, shown by straight lines for N=1, 2, 3 and 4, are adduced. In the same figure, the dependences of ratio of sum $(\varphi_n+\varphi_{if})$, which in the model [2] is f_p analogue, and W_n, obtained according to the equations (6) and (7), for nanocomposites EP/MMT (points) are given. The similarity of trends of the indicated ratios as a function of d_{001}/d_{pl} change is observed again, which confirms again "effective particle" definition in the models [2] and [3] identity. And as earlier, for nanocomposites EP/MMT, the value N can be estimated as somewhat smaller than N=2.

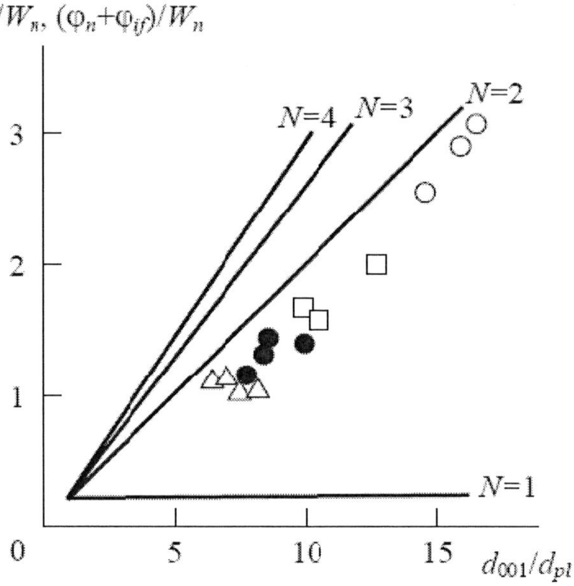

Figure 4. The dependences of organoclay fraction in "effective particles" volume and mass fractions ratio f_p/W_n in the model [3] for silicate platelets number in it N=1 (1), 2 (2), 3 (3) and 4 (4) and $(\varphi_n+\varphi_{if})/W_n$ in the model [2] for nanocomposites EP/MMT (points) on the ratio d_{001}/d_{pl} value. Conventional signs are the same as in Figure 2.

CONCLUSION

As the present paper results have shown, the model of organoclay "effective particle" offered by the authors [2] describes nanocomposites polymer/organoclay structure and properties so well, as the multiscale micromechanical model, elaborated in the paper [3]. However, unlike the latter, the model [2] is much simpler, has clear physical significance, does not require micromechanical models application and, hence, organoclay and interfacial regions characteristics usage, but accounts for interfacial adhesion polymeric matrix – nanofiller real level. The indicated factors make the model [2] more suitable for applied calculations and clearer from the point of view of physical treatment in comparison with the model [3].

REFERENCES

[1] Malamatov, A.Kh., Kozlov, G.V., Mikitaev, M.A. The Reinforcement Mechanisms of Polymer Nanocomposites. Moscow, Publishers the D.I. Mendeleev RKhTU, 2006, 240 p.

[2] Mikitaev, A.K., Kozlov, G.V., Zaikov, G.E. Polymer Nanocomposites: Variety of Structural Forms and Applications. New York, Nova Science Publishers, Inc., 2008, 319 p.

[3] Sheng, N., Boyce, M.C., Parks, D.M., Rutledge, G.C., Abes, J.I., Cohen, R.E. Multiscale micromechanical modeling of polymer/clay nanocomposites and the effective clay particle. *Polymer,* 2004, v. 45, № 2, pp. 487-506.

[4] Ahmed, S., Jones, F.R. A review of particulate reinforcement theories of polymer composites. *J. Mater. Sci.,* 1990, v. 25, № 12, pp. 4933-4942.

[5] Bobryshev, A.N., Kozomazov, V.N., Babin, L.O., Solomatov, V.I. Synergetics of Composite Materials. Lipetsk, NPO ORIUS, 1994, 154 p.

[6] Novikov, V.U., Kozlov, G.V. The fractal parametrization of filled polymers structure. *Mekhanika Kompozitnykh Materialov*, 1999, v. 35, № 3, pp. 269-290.

[7] Chen, J.-S., Poliks, M.D., Ober, C.K., Zhang, Y., Wiesner, U., Giannelis, E. Study of the interlayer expansion mechanism and thermal-mechanical properties of surface-initiated epoxy nanocomposites. *Polymer,* 2002, v. 43, № 14, pp. 4895-4904.

[8] Kozlov, G.V., Malamatov, A.Kh., Antipov, E.M., Karnet, Yu.N., Yanovskii, Yu.G. Structure and mechanical properties of polymer nanocomposites within the frameworks of fractal concepts. *Mekhanika Kompozitsionnykh Materialov i Konstruktsii,* 2006, v. 12, № 1, pp. 99-140.

[9] Kozlov, G.V., Aphashagova, Z.Kh., Burya, A.I., Lipatov, Yu.S. Nanoadhesion and reinforcement mechanism of particulate-filled polymer nanocomposites. *Inzhenernaya Fizika,* 2008, № 1, pp. 47-50.

[10] Dennis, H.R., Hunter, D.L., Chang, D., Kim, S., White, J.L., Cho, J.W., Paul, D.R. Effect of melt processing conditions on extent of exfoliation in organoclay-based nanocomposites. *Polymer,* 2001, v. 42, № 24, pp. 9513-9522.

[11] Pernyeszi, T., Dekany, I. Surface fractal and structural properties of layered clay minerals monitored by small-angle X-ray scattering and low-temperature nitrogen adsorption experiments. *Colloid Polymer Sci.,* 2003, v. 281, № 1, pp. 73-78.

[12] Kubat, J., Rigdahl, M., Welander, M. Characterization of interfacial interactions in high density polyethylene filled with glass spheres using dynamic-mechanical analysis. *J. Appl. Polymer Sci.*, 1990, v. 39, № 9, pp. 1527-1539.

In: Development of Nanotechnology in Textiles ISBN: 978-1-62081-030-9
Editors: A. K. Haghi and G. E. Zaikov ©2012 Nova Science Publishers, Inc.

Chapter 6

ECO-FLAME RETARDANTS

A. M. Sakharov, P. A. Sakharov, S. M. Lomakin and G. E. Zaikov
N.M. Emanuel's Institute of Biochemical Physics
Russian Academy of Sciences,
Moscow, Russia

INTRODUCTION

Today, as well a lot of centuries back, the prevention and extinguishing of fires is one of the global problems standing in front of humanity. Throughout the world annually, thousands of new reagents appear directed on the decision of this problem. But as we can see, the problems such as the extinguishing of forest fires or creation of low inflammable polymeric materials is far from the solution.

The combustion of natural and polymer materials, like the combustion process of any other fuel material, is a combination of complex physical and chemical processes, which includes the transformation of initial products. This whole conversion process may be divided into stages, with specific physical and chemical processes occurring in each of these stages. In contrast to the combustion of gases, the combustion process of condensed substances has a multi-phase character. Each stage of the initial transformation of a substance correlates with a corresponding value (combustion wave) with specific physical and chemical properties (state of aggregation, temperature range, concentration of the reacting substances, kinetic parameters of the reaction etc.).

Char-forming materials often swell and intumesce during their degradation (combustion), and the flame-retardant approach is to promote the formation of such intumescent char. The study of new polymer flame retardants has been directed at finding ways to increase the [complete text]

There is a strong correlation between char yield and fire resistance. This follows because char is formed at the expense of combustible gases and because the presence of a char inhibits further flame spread by acting as a thermal barrier around the unburned material. The tendency of a polymer to char can be increased with chemical additives and by altering its molecular structure. Polymeric additives (polyvinyl alcohol systems) which promote the formation of char have been studied [3-5].

These polymeric additives usually produce a highly conjugated system - aromatic structures which char during thermal degradation and/or transform into cross-linking agents at high temperatures.

Char, which is formed in the process of thermal degradation, can play several roles in fire retardancy. The formation of char in and of itself has a significant effect on the degradation because char formation must occur at the expense of other reactions that may form volatiles; thus, char formation may limit the amount of fuel available. An example of this occurs in cellulose, which may degrade either by a series of dehydration reactions that yield water, carbon dioxide, and char, or by a process in which levoglucosan is produced, which eventually leads to the formation of volatiles.

It is believed that the temperature at the surface of a burning polymer is close to the temperature at which extensive thermal degradation occurs (usually 300-600°C). The bottom layer of char, near the polymer surface, is at the same temperature, whereas the upper surface, exposed to the flame, can be as hot as 1500°C [6].

Therefore, fire-retardancy chemistry is concerned with chars, which may be produced at temperatures between 300° and 1500°C [7].

Char-forming materials often swell and intumesce during their degradation (combustion), and the flame-retardant approach is to promote the formation of such intumescent char. Intumescent systems, on heating, give a swollen multicellular char capable of protecting the underlying material from the action of the flame. Fire-protective coatings with intumescent properties have been in use for 50 years, whereas the incorporation of intumescent additives in polymeric materials is a relatively recent approach [8-10].

On burning, these additives develop the foamed char on the surface of degraded material. The suggested mechanism of fire retardancy assumes that the char acts as a physical barrier against heat transmission and diffusion of oxygen toward the polymer and of combustible degradation products of the polymer toward the flame. Thus, the rate of pyrolysis of the natural and polymeric materials is expected to decrease below flame-feeding requirements, which leads to flame extinguishment. The intumescent behavior resulting from a combination of charring and foaming of the surface of the burning polymers is being widely developed for fire retardancy because it is characterized by a low environmental impact.

Apparently, intumescent coatings are the most effective methods of protecting wood and natural materials from fire. When heated, they form a thick, porous carbonaceous layer. This provides an ideal insulation of the protected surface against an excessive increase in temperature and oxygen availability, thus preventing thermal decomposition which plays a decisive role in retarding the combustion process. Under the heating, intumescent compounds expand up to 200 times their volume, and, in some cases, to even 200 times [11]. [Vandersall H L, Intumescent coating systems, their development and chemistry *J Fire Flamm*, 1971, 2 April: 97–140]. Intumescent systems such as paints, lacquers, mastics and linings must contain ingredients which, when heated to high temperatures, will form large amounts of non-flammable residues. These residues, under the influence of emitted gases, produce foam with good insulating properties. [11,12,13].

Heating of cellulosic materials and wood at 105 -110°C results in the removal of moisture. Reactions occurring at this stage proceed slowly and are mostly endothermic. After exceeding the above temperature, a slow thermal decomposition of the components of natural polymers begins and at 150–200°C, gas products of the decomposition start to be released.

At 160°C, decomposition of lignin begins. Lignin under thermal degradation yields phenols from cleavage of ether and carbon–carbon linkages, resulting in more char than in the case of cellulose. Most of the fixed carbon in charcoal originates from lignin. At 180°C, hydrolysis of low molecular weight polysaccharides (hemicelluloses) begins [13]. Thermal stability of hemicelluloses is lower than that of cellulose and they release more incombustible gases and fewer tarry substances [12] The most abundant gaseous products contain about 70% of incombustible CO_2 and about 30% of combustible CO. Depending on the availability of oxygen, subsequent reactions can be exothermic or endothermic. In the temperature range of 220–260°C, exothermic reactions begin. They are characterized by an evolution of gaseous products of decomposition, release of tarry substances and the appearance of ignition areas of hydrocarbons with low boiling points. Cellulose decomposes in the temperature range between 260 and 350°C, and it is primarily responsible for the formation of flammable volatile compounds. [12].

At temperatures of about 275–280°C, the accelerated release of considerable quantities of heat begins and increased amounts of liquid and gaseous products (280–300°C), including methanol, acetic acid are formed. The amount of evolving carbon monoxide and dioxide decreases, mechanical slackening of wood structure proceeds and ignition occurs. Mass loss of wood reaches about 39%. [12] Tar begins to appear at 290°C. The release of gases still increases and a rapid formation of charcoal takes place. This reaction is highly exothermic and proceeds at 280–320°C. Secondary reactions of pyrolysis become predominant resulting in an increased amount of gaseous products. Combustion proceeds in the gas phase at a small distance from the surface rather than on the wood surface itself. The ignition of wood occurs at 300–400°C depending on its origin and lignin content [13].

At the final stage of combustion of wood (above 500°C), the formation of combustible compounds is small and the formation of charcoal increases. Charcoal makes an insulating layer which hinders heat transfer, thus preventing the temperature from increasing to the pyrolysis of wood occurs.

Mechanisms of flame retardancy depend on the character of the action of the flame retardants chemical compounds present in fire retardants. In a comprehensive review on wood flame retardants, R. Kozlowski and M. Wladyka-Przybylak proposed two general groups of fire retardants for wood: additive and reactive [12]. Additive compounds are those whose interaction with a substrate is only physical in its nature, whereas reactive compounds interact chemically with cellulose, hemicellulose or lignin [12, 14]. The applied additive compounds include: mono- and diammonium phosphate, halogenated phosphate esters, phosphonates, inorganic compounds such as antimony oxide and halogens, ammonium salts (ammonium bromide, ammonium fluoroborate, ammonium polyphosphate and ammonium chloride); amino resins (compounds used for their manufacture are dicyandiamide, phosphoric acid, formaldehyde, melamine and urea); hydrated alumina; stannic oxide hydrate; zinc chloride and boron compounds (boric acid, borax, zinc borate, triammonium borate, ethyl and methyl borates) [12].

One of the most effective methods of protecting wood from fire is the use of fire-retardant intumescent coatings. When heated, they form a thick, porous carbonaceous layer. This provides an ideal insulation of the protected surface against an excessive increase in temperature and atmospheric oxygen, thus preventing thermal decomposition which plays a decisive role in retarding the combustion process.

Studies of new flame-retardant systems for the protection of natural polymers and wood can be directed to chemical modification of natural polymers and wood and the use of more efficient intumescent systems and fire protectors. Here we represent the results of researches of use of non-polluting reagents from renewed vegetative raw materials as highly effective flame retardants intumescent/char-forming systems for cellulosic materials and wood which can find wide applications due to their cheapness and simple technology of production.

Earlier it has been found that starch, cellulose and lignin can be easy oxidized by oxygen in the presence of copper salts and alkali. The basic products of oxidation of polysaccharides and lignin are salts of polyoxyacids [15]. Such salts can be used as components of washing powders, components of boring solutions, gums for use in building and etc. But unexpectedly, it was found that the salts of polyoxyacids are also diminished in the burning of different materials [16]. In the present work, the data on properties of the oxidized polysaccharides and a lignin, used as flame retardants is presented.

Oxidized Polysaccharides and Lignin

The methods of oxidative modifying of polysaccharides with receiving of polyacids as main products are widely used in practice due to the availability of initial raw material and high consumer properties of oxidized polysaccharides. Salts of polyacids are widely used as watersoluble glues in the production of paper, cardboard, in the processes of materials dressing, as components of drilling agents, etc. However, as well as for a lot of other processes of organic compounds oxidation as an oxidizing agent of polysaccharides, the hypochlorites and periodates

[17,18], hydrogen peroxide and not gaseous oxygen are used until recently that is connected with low activity of oxygen in processes of polysaccharides oxidation.

In the presence of copper complexes and bases not only simple in their structure alcohols and ketones, but also polysaccharides (starches, dextranes, cellulose) may be oxidized by oxygen with high rates. High rates of polysaccharides oxidation by oxygen exceeding oxidation rates by hypochlorites and other oxidizing agents are reached at temperatures 40-90°C [19].

Oxidation of polysaccharides with the greatest efficiency proceeds in the presence of copper and alkali salts. In Figure 1, kinetic curve of oxygen absorption (curve 1), NaOH consumption (curve 2) and changes of viscosity of initial gel (curve 3) during 10% of potato starch gel oxidation are presented. As seen from Figure1 at the first 10-20 minutes of starch oxidation, viscosity of initial gel decreases a hundred times and final (after 3-4 hours) viscosity of a solution comes near to the viscosity of water. The molecular weight of an initial polysaccharide in the first minutes decreases slightly [20].

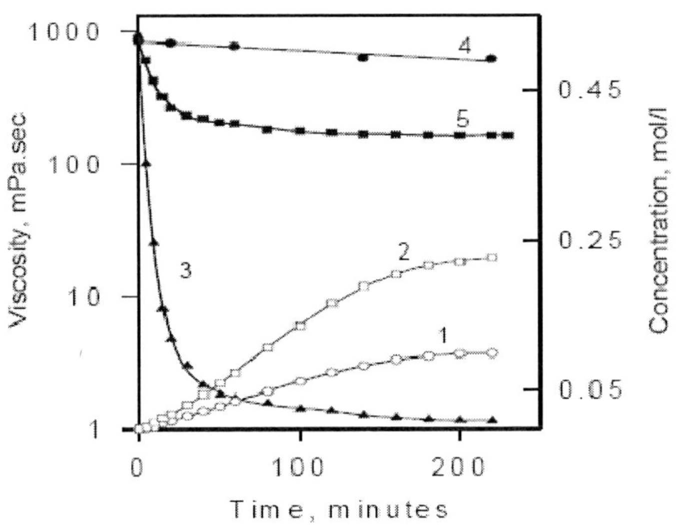

Figure 1. The kinetic curves of O_2 (1) and NaOH (2) consumption and change of viscosity of initial gel in the course of starch oxidation in the presents of copper ions (3) and change of viscosity of gel in the absence of catalyst (4). Curve 5 – change of the viscosity of initial gel of starch (5 g starch and 0.5 g NaOH in 50 ml H_2O) under inert atmosphere after addition of 1g dry powder of deep oxidized starch. $[CuCl_2]_o$= 5.10^{-3} M , 5g potato starch, 0.5g NaOH, 50 ml H_2O, 338 K.

Without copper salt, the oxidation of starch does not proceed with a measureable rate and viscosity of gel practically does not change (curve 4). But an addition of oxidized starch to the native gel of starch without oxygen drastically diminishes the viscosity of the initial gel (curve 5).

Below, the most probable formula of the oxidized polysaccharide (starch) is presented. The chemical formula of the oxidized polysaccharides includes unmodified α-D-glucopyranosyl cycles and the oxidized links of polysaccharides containing carboxyl groups [20].

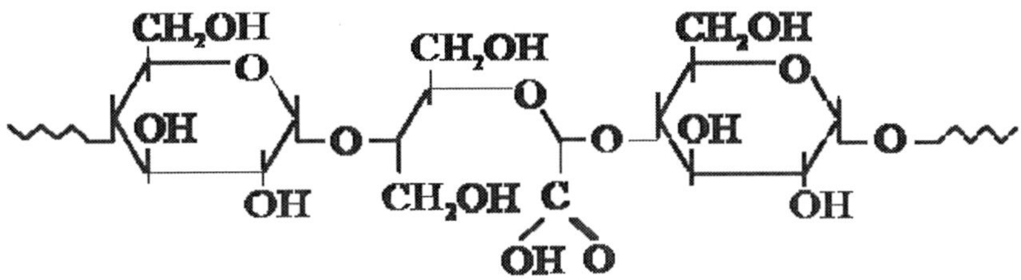

Copper salts are the most effective catalysts of oxidation of polysaccharides in the presence of alkali. For lack of the catalyst, starch and cellulose practically are not oxidized at temperatures below 373 K. However, lignin which contains many of high reactive phenolic groups (under high rates), is oxidized by oxygen and without the catalyst with measurable rates.

As well as under oxidation of polyols with low molecular weights [21], the anion forms of polysaccharides reacts with Cu^{2+} ions with $[Cu^{2+}...A^{-}]$ adducts formation (A^{-} – deprotonated polysaccharide). The function of the catalyst (Cu^{2+} ions) is to activate deprotonated forms of substrate to oxygen. It is possible to suppose that the role of bivalent copper ions is the oxidation of the anion form of substrate and followed interaction of intermediate radicals or anion radicals with O_2. However, high-molecular polysaccharides with a low content of end aldehyde groups in anaerobic medium are redox inactive and the rate of reduction of Cu^{2} to Cu^{+}- ions (due to electron transfer from substrate anion form to Cu^{2+}) proceeds with rates in a hundred times lower than the rate of oxygen adsorption during the process of polysaccharides oxidation in alkaline mediums. Apparently, just direct interaction of oxygen with anions A^{-} in coordination sphere of copper ion leads to the formation of hydroxycarboxylates as the main primary products of oxidation. Absorption of oxygen is completely stagnated after neutralization of introduced alkali by polyoxyacids formed during the oxidation process. Thus, by varying the amountof introduced alkali, we may change the degree of polysaccharides oxidation into polyacids. This fact allows us to change the final products viscosity, bonding ability, solubility in water, etc.

As initial raw materials for receiving of salts of polyoxyacids, not only starch can be used, but any other starch-containing raw material: corns of maize, oats, rice, etc., including ill-conditioned raw materials (grain-crops affected by various fungus diseases, waste of rice slashing, waste of mill houses, etc.). This fact enables to decrease the prime cost of the final product significantly. Moreover, the similarity to the polysaccharides behavior of lignin under oxidation conditions was observed in alkali media with/without the catalyst. Lignin is among the most abundant biopolymers on earth. It is estimated that the planet currently contains 3 x 10^{11} metric tons of lignin with an annual biosynthetic rate of approximately 2 x 10^{10} tons [22]. Lignin constitutes approximately 30% of the dry weight of softwoods and about 20% of the weight of hardwoods [23]. Lignin is absent from primitive plants such as algae, and fungi which lack a vascular system and mechanical reinforcement. The presence of lignin within the cellulosic fiber wall, mixed with hemicelluloses, creates a naturally occurring composite material which imparts strength and rigidity to trees and plants.

Lignin is a random co-polymer consisting of phenylpropane units having characteristic side chains. Lignin slightly cross-links and takes an amorphous structure in the solid state. The molecular motion is observed as glass transition by thermal, viscoelastic and spectroscopic measurements. The hydroxyl group of lignin plays a crucial role in interaction with water. Lignin is usually considered as a polyphenolic material having an amorphous structure, which arises from an enzyme-initiated dehydrogenative polymerization of *p*-coumaryl, coniferyl and sinapyl alcohols. The basic lignin structure is classified into only two components; one is the aromatic part and the other is the C3 chain. The only usable reaction site in lignin is the OH group, which is the case for both phenolic and alcoholic hydroxyl groups.

Lignin is one of the most important bio-resources for the raw material of the synthesis of environmentally compatible polymers. Lignins are derived from renewable resources such as trees, grasses, and agricultural crops. About 30% of wood constituents are lignin. Lignins are non-toxic and extremely versatile in performance. Production of lignin as a by-product of pulping process in the world is over 30 million tons per year [23].

According to the widely accepted concept, lignin may be defined as an amorphous, three-dimensional polyphenolic polymer arising from an enzyme-mediated dehydrogenative polymerization of three phenylpropanoid monomers, coniferyl, sinapyl, and p-coumaryl alcohols. Basically, three major structures of lignin, 4-hydroxyphenyl (1), guaiacyl (2), and syringyl (3) structures are conjugated to produce a lignin polymer in the process of radical-based lignin biosynthesis.

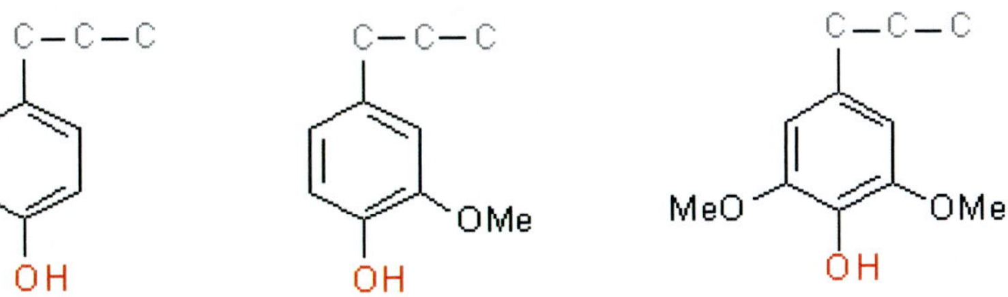

(1) (2) (3)

Results of estimation of efficiency of fire protecting action of coverings based on oxidized starch-containing raw materials with different amounts of applied reagent were carried out by standard method (ASTM E136-09 Standard Test Method for Behavior of Materials in a Vertical Tube Furnace) (Figure 2).

It is obvious that all pine samples treated with "one-layer covering" of oxidized starch reagent (OSR) - 100 g/m^2 provide good fire-protection activity (II group) to reach a rank of hardly-inflammable wood. In the case of "multi-layer" covering, 3÷4 covering, (OSR consumption 300÷400 g/m^2) mass losses at fire testing are significantly decreased, allowing us to obtain the I group of fire-protection – hardly-inflammable wood.

Mode of Action of Oxidized Polysaccharides (Intumescence Behavior)

Study of thermal decomposition of oxidized starch at dynamic heating in the range of temperatures 25 – 950 °C has shown that formation of the foam coke occurs at the very first stage at temperature 150 – 280 °C as a result of synchronous processes of reduction of viscosity of polymer at its transition from glass forming form to viscous-fluid and chemical reactions of decarboxylation and dehydration. Making foam agents are gaseous products of decomposition – water and carbon dioxide.

Reactions of intermolecular dehydration promote the formation of the sewed spatially-mesh structure and stabilization and hardening of thefoam coke. Figure2 shows the formation of intumescent coke from oxidized starch under the heating at 400°C in air.

Figure 2. Formations of coke after heat treatment of a metal plate with covering oxidized starch.

The combustibility study of pinewood samples treated with aqueous solution of oxidized starch have shown a high efficiency of flame retardant action of this intumescent additive during combustion process.

Fire tests results of plain pinewood samples and the pine samples treated with aqueous solution of oxidized starch according to ASTM E136-09 (Standard Test Method for Behavior of Materials in a Vertical Tube Furnace) (Figs. 3, 4).

Figure 3. Photograph of initial pinewood samples before fire test.

The effect of fireproof action of high-molecular oxidized starch flame retardant is connected with formation on a surface of a wood-sufficient foam coke layer. This foam coke shows excellent heat-shielding barrier properties, complicates heat access to the wood surface and hinders the evolution of combustible gaseous products of the decomposition of wood.

Figure 4. Photograph of pinewood samples treated with aqueous solution oxidized starch flame retardant after fire tests (ASTM E136-09).

Figure 5 shows the actual mass loss of pinewood samples treated with aqueous solutions of different oxidized polysaccharides after fire tests (ASTM E136-09).

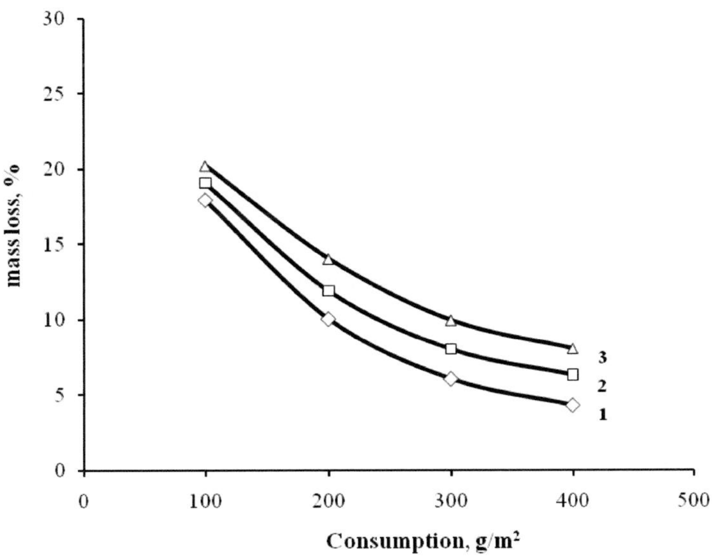

Figure 5. Dependence of mass losses of wood samples on amount of applied fire-protecting coverings based on modified polysaccharides: 1 – oxidized starch (high degree of oxidation), 2 – oxidized rice (high degree of oxidation), 3 – oxidized starch (average degree of oxidation).

The combustibility of pristine pinewood samples and those treated by oxidized lignin and starch were also evaluated by the Mass Loss Calorimeter (ISO 13927) at an incident heat flux of 35 kW/m^2 condition [24]. Mass loss rates (MLR) of pinewood samples acquired at 35 kW/m^2 are presented in Figure 6.

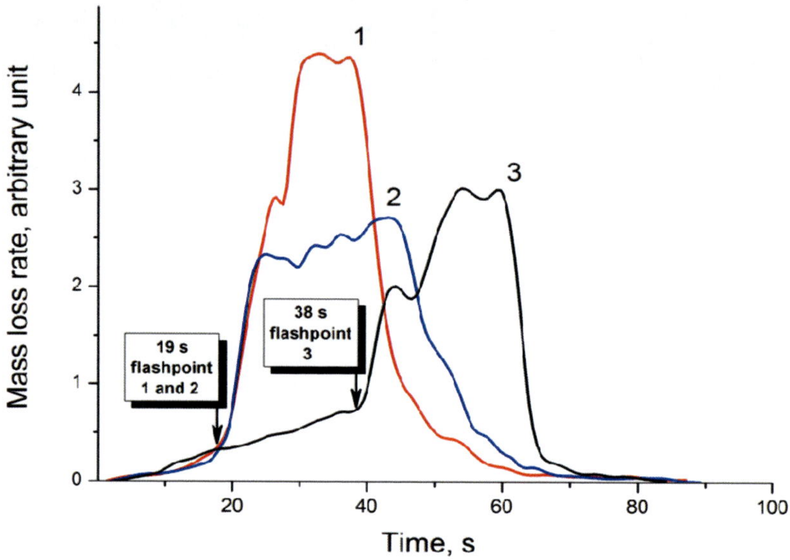

Figure 6. Mass loss rate vs. time for plain pinewood sample - (1), pinewood sample treated with of oxidized lignin (15% by wt.) - (2) and pinewood sample treated with oxidized starch (15% by wt.) – (3).

It is clearly seen that under conditions of initial surface combustion, the mass loss rate of the pinewood samples treated with of oxidized lignin and starch (15% by wt.) and the pinewood sample treated with oxidized starch and its rate are noticeably lower than an adequate value for the neat pinewood samples. The peak of MLR of the pinewood sample treated with of oxidized lignin is 41% lower than that of pure pinewood sample, while that of the pinewood sample treated with oxidized starch is 33% lower. Due to predominant solid-phase mode of action of oxidized lignin and starch flame retardants, the MLR curves are supposed to be similar to the RHR (rate of heat release) curves, so the reduction of the MLR is evidently the primary factor responsible for the lower RHR of the pinewood samples. An improvement in flame resistance of the pinewood samples treated with of oxidized lignin over the pinewood sample happens as a result of the char formation providing a transient protective barrier of oxidized lignin (as polyphenolic carbonizing structure), whereas the pinewood sample treated with oxidized starch indicates the strong intumescent action. An additional observation obtained from the MLR plots is the substantial increase in the induction period time of self-ignition (flashpoint) of the pinewood sample treated with oxidized starch as compared with plain pinewood sample and the pinewood sample treated with oxidized lignin due to intumescent behavior of oxidized starch (Figure6).

REFERENCES

[1] Lomakin, S.M., Brown, J.E., Breese, R.S., Nyden, M.R. *Polymer Degrad. and Stab.* 41, 229-243 (1993).

[2] Factor, A. Char Formation in Aromatic Engineering Polymers, in: Fire and Polymers, Ed. Nelson, G.L. (Ed.), pp. 274-287, *ACS Symposium Series* 425, ASC, Washington DC (1990).

[3] Lomakin, S.M., Zaikov, G.E., Artsis M.I. *Intern. J. Polym. Mater.* 32 (1-4) 173-202 (1996).

[4] Lomakin, S.M., Zaikov, G.E., Artsis, M.I., Ruban, L.V., Aseeva R.M. *Oxidation Comm.* 18 (2), 105-112 (1995).

[5] Lomakin, S.M., Zaikov, G.E., Artsis M.I. *Intern. J. Polym. Mater.* 26 (3-4), 187-194 (1994).

[6] Aseeva, R.M., Zaikov, G.E. *Combustion of Polymeric Materials.* Moscow: Nauka, 1981, pp. 84-135 (in Russian).

[7] Levchik, S., Wilkie, Ch.A. Char Formation, Chapter 6 in: Fire Retardancy of Polymeric Materials, ed. By Grand, A.F., Wilkie, Ch.A. Marcel Dekker, Inc. New York (2000), 171-216

[8] Camino, G. Flame retardants: Intumescent systems. In: Pritchard, G. ed. *Plastics Additives,* London: Chapman Hall, 1998, pp. 297-306.

[9] Camino,G., Delobel, R. Intumescence, Chapter 7 in Fire Retardancy of Polymeric Materials, ed. By Grand, A.F., Wilkie, Ch.A. Marcel Dekker, Inc. New York (2000), 217-244

[10] Camino, G., Lomakin, S. Intumescent materials In: Horrocks, A.R., Price, D., eds., Fire Retardant Materials, CRC Press, New York, Washington DC (2001), 318-335.

[11] Vandersall, H.L. Intumescent coating systems, their development and chemistry. *J. Fire Flamm.* 1971, 2 April: 97–140

[12] Kozlowski, R., Wladyka-Przybylak, M. Natural polymers, wood and lignocellulosic materials In: Horrocks, A.R., Price, D., eds., *Fire*

Retardant Materials, CRC Press, New York, Washington DC (2001), 293-317.

[13] Hurst, N.W., Jones, T.A. A review of products evolved from heated coal, wood and PVC, *Fire Mater,* 1985, 9(1), 1–9.

[14] Lewin, M., Atlas, S.M., Pearce, E.M. *'Flame-Retardant Polymeric Materials',* vol 1, New York and London, Plenum Press, 1975.

[15] Sakharov, A.M. In: *Chemical and Biochemical Physics: New Frontiers* (2006), Editor: Zaikov, G.E., Nova Science Publishers, 113.

[16] Sivenkov, A.B., Serkov, B.B., Aseeva, R.M., Sakharov, A.M., Sakharov, P.A., Skibida, I.P. (2002). *Fire & Explosion Safety.* 1, 39.(in Russian).

[17] Floor, M., Kieboom, A.P.G., van Bekhum, H. (1989). *Starch,* 41, 303.

[18] Santacesria, E., Trully, F., Brussani, G.F., Gelosa, D., Serio, M.Di. (1994).

[19] Skibida I.P., Sakharov An.M., Sakharov Al.M. (1993). EP 91122164.6, *Carbohydrate Polymers,* 23, 35.

[20] Sakharov, A.M., Skibida, I.P. (2001). *Chem. Phys.,* 20, 101.

[21] Sakharov, A.M., Silakhtaryan, N.T., Skibida, I.P. (1996). *Kinetics and Catalysis,* 37, 393.

[22] Cereal Straw as a Resource for Sustainable Biomaterials and Biofuels: Chemistry, Extractives, Lignins, Hemicelluloses and Cellulose, Ed. By Run-Cang Sun Elsevier, 2010, 292 p.

[23] Hatakeyama, H., Hatakeyama, T. Lignin Structure, Properties, and Applications. In Biopolymers - Lignin, Proteins, Bioactive Nanocomposites, ed. By Abe, A., Dušek, K.,Kobayashi, Sh. Springer, Heidelberg, Dordrecht, London, New York, 2010, p.2-64

[24] Standard Test Method for Screening Test for Mass Loss and Ignitability of Materials (ASTM E2102). American Society for Testing and Materials, Philadelphia, PA.

In: Development of Nanotechnology in Textiles ISBN: 978-1-62081-030-9
Editors: A. K. Haghi and G. E. Zaikov ©2012 Nova Science Publishers, Inc.

Chapter 7

SYNTHESIS AND CHARACTERIZATION OF A HYPERBRANCHED POLYMER

M. Hasanzadeh and M. Haghighat Kish
Department of Textile Engineering,
Amirkabir University of Technology, Tehran, Iran

1. INTRODUCTION

A novel hyperbranched, functional, water-soluble and amine terminated polymer was synthesized by melt polycondensation reaction of methyl acrylate and diethylene triamine. The hyperbranched polymer sample prepared was characterized by Fourier transform infrared (FTIR) and nuclear magnetic resonance (NMR) spectroscopy. Poly(ethylene terephthalate) (PET) fabrics and hydrolyzed PET fabrics with sodium hydroxide were treated with the synthesized hyperbranched polymer. Color strength of hyperbranched polymer-treated PET fabrics exhibited improvements when dyed with acid dye (C.I. Acid Red 114) in conventional dyeing system. The color characteristics of the dyed fabrics were evaluated using CIELAB method. The results of the measurement on the fabrics after competitive dyeing with acid dyes indicated that the hyperbranched polymer, as indicated by zeta potential measurement, offer positively charged amino groups for differential-dyeing. The treated samples exhibited good wash fastness. The resulting hyperbranched polymer is readily soluble in polar solvents such as water, DMSO and DMF.

To optimize and predict the dyeability of PET fabrics which treated with hyperbranched polymer, CCD experimental design was employed. A more systematic understanding of process parameters was obtained and a quantitative relationship between treatment conditions and fabrics dyeability was established by response surface methodology (RSM).

In recent years, dendritic polymers have attracted increasing attention due to their unique chemical and physical properties. These polymers consist of three subsets, namely hyperbranched polymers, dendrigraft polymers and dendrimers. Hyperbranched polymers (HBPs) are highly branched, polydisperse and three-dimensional macromolecules synthesized from a multifunctional monomer to produce a molecule with dendritic structure [1-10].

Dendrimers are well-defined and need a stepwise route to construct the perfectly symmetrical structure. Hence, synthesis of dendrimers is time-consuming and expensive procedures. Although hyperbranched polymers are irregularly shaped and not perfectly symmetrical like dendrimers, hyperbranched polymers rapidly prepared and generally synthesized by one-step process via polyaddition, polycondensation, radical polymerization, and so on, of AB_x type monomers and no purification steps are needed for their preparation. Therefore, HBPs are attractive materials for industrial applications due to their simple production process [1-4, 11-19]. In general, according to molecular structures and properties, hyperbranched polymers represent a transition between linear polymers and perfect dendrimers. Comparison of hyperbranched polymers with their linear analogues indicated that HBPs have remarkable properties, such as low melt and solution viscosity, low chain entanglement, and high solubility, as a result of the large amount of functional end groups and globular structure [10-16].

In recent years, many functional hyperbranched polymers with various terminal groups such as hydroxyl, amine, carboxyl, acetoxy, and vinyl have been suggested as excellent candidate for use in drug delivery [20, 21], gene therapy vectors [21, 22], coatings [23,24], additives [25], catalysis [26], gas separation [27], nanotechnology and supramolecular science [28], and many more.

Some examples of modification of fibers with hyperbranched polymers were successful. For instance, the dyeability of modified polypropylene (PP) fibers by hyperbranched polymer was investigated [29]. The results showed that the incorporation of hyperbranched polymer prior to fiber spinning considerably improved the color strength of polypropylene fiber with C.I.

Disperse Blue 56 and has no significant effect on physical properties of the PP fibers.

The synthesized amino-terminated hyperbranched polymer, was found to be used as salt-free dyeing auxiliary for treated cotton fabrics [30]. The washing fastness, rubbing fastness and leveling properties of hyperbranched polymer treated cotton fabrics were better than untreated cotton fabrics. In the study on applying amino-terminated hyperbranched polymer to cotton fabric, it was demonstrated that hyperbranched polymer treatment on cotton fabrics has no undesirable effect on mechanical properties of fabrics. Dyeing of treated cotton fabrics with direct and reactive dyes in the absence of electrolyte showed that the color strength of treated samples were better than untreated cotton fabrics. Furthermore, application of hyperbranched polymer to cotton fabrics reduced UV transmission and showed good antibacterial activities [31-34].

In recent years, there has been considerable research to improve poly(ethylene terephthalate) (PET) fabrics dyeability. It is known that PET fabrics are usually dyed using disperse dyes in the presence of a carrier or at elevated temperatures. Many attempts have been made during the last two decades to replace environmentally unfriendly carriers with non-toxic chemicals [35]. Modification of PET fabrics by attaching additives or functional groups to the polymer molecules via grafting and co-polymerization are the common methods to enhance the dyeability of PET fabrics.

Literature review showed that there has not been a previous report regarding the treatment of amine-terminated hyperbranched polymers on PET fabric and study of its dyeability with acid dyes. In the most recent investigation in this field, fiber grade PET was compounded with polyesteramide hyperbranched polymer and dyeability of resulted samples with disperse dyes was studied [35]. The results showed that the dyeability of dyed modified samples comprised of fiber grade PET films and a hyperbranched polymer (Hybrane H1500) were better than the neat PET and this was increased by an increasing amount of hyperbranched polymer in the presence or absence of a carrier. The dyeability of the samples was attributed to decrease in glass transition temperature for blended PET/hyperbranched polymer in comparison with neat PET [29].

In the first part of our present work, novel hyperbranched polymer (HBP) with amine terminal group was synthesized from methyl acrylate and diethylene triamine by melt-polycondensation. The obtained hyperbranched polymer was characterized using Fourier transform infrared spectroscopy (FTIR) and nuclear magnetic resonance (NMR). In this contribution for the first time, the synthesized hyperbranched polymer applied to PET fabric and the dyeability of HBP-treated PET fabrics and untreated PET fabric with C.I. Acid Red 114 was investigated. Dyeing of PET fabric with acid dye is very helpful and attractive object.

In the second part, a different case study was presented to show how the treatment conditions can be affected the dyeability of PET fabrics. Response surface methodology (RSM) was used to obtain a quantitative relationship between selected treatment conditions and PET fabrics dyeabiliy.

2. EXPERIMENTAL

2.1. Case 1- Synthesis, Characterization, and Application of HBP to PET Fabrics

2.1.1. Materials

Methyl acrylate, diethylene triamine of pure grade were obtained from Merck, Germany. All reagents used were analytically pure. Poly(ethylene terephthalate) (PET) fabric (28×19 count/cm^2) used throughout this work and before use was treated with a solution containing 5 g/L Na_2CO_3 and 1 g/L of a non-ionic detergent at 60 °C for 30 min to remove undesired materials. Distilled water was used for the treatments and washings. Acid dye (C.I. Acid Red 114) was provided by the Ciba Ltd. (Tehran, Iran) and used to evaluate the dye absorption behavior (dyeability) of samples. The chemical structure of AR 114 is shown in Figure 1.

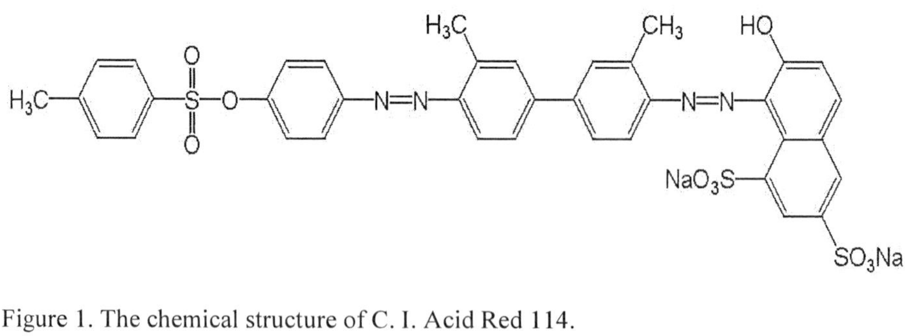

Figure 1. The chemical structure of C. I. Acid Red 114.

2.1.2. Measurements

The infrared spectra of samples were recorded by Nicolet 670 FTIR spectrophotometer in the wave number range 500–4000 cm^{-1}. Nominal resolution for all spectra was $4cm^{-1}$.

The ^{1}H and ^{13}C NMR spectra were recorded by Bruker AVANCE (500MHz for ^{1}H and 125MHz for ^{13}C) using DMSO-d_6 as a solvent at room temperature. All chemical shifts (δ) are expressed in ppm, the following abbreviations are used to explain the multiplicities: s = singlet, d = doublet, t = triplet, q = quartet, m = multiplet.

X-ray diffraction (XRD) patterns of hyperbranched-treated and untreated polyester fabric were carried out with Equinox 3000 X-ray diffractometer with Cu Kα radiation (λ=0.154 nm) and operated at 40 kV voltage and 30 mA electric current.

The ultraviolet–visible (UV–vis) absorption spectrum of the hyperbranched polymer was taken with a Carry 100 UV–vis spectrometer.

The zeta potential of the untreated polyester fabric and the HBP-treated polyester fabrics were measured by a Molvern Zetasizer 3600 device. Fabric suspension prepared by pulling out the fibers from the fabric and cutting the fibers into lengths of approximately 0.5 mm. Then the small piece of fibers soaked in deionized water for 24 h.

The color parameters of dyed samples were obtained under illuminant D_{65} at 10° standard observer in the visible range using Color-Eye 7000A spectrophotometer. The Kubelka–Munk single constant theory was used to calculate K/S values at the wavelength of maximum absorption (λ_{max}) for each fabric as the following equation:

$$\frac{K}{S} = \frac{(1-R)^2}{2R} - \frac{(1-R_0)^2}{2R_0} \qquad (1)$$

where R is the reflectance value at the wavelengths of maximum absorbance (λ_{max}), K is the absorption coefficient and S is the scattering coefficient.

Washing fastness test was performed according to ISO/R 105/IV, Part 10. Rubbing fastness test was performed according to ISO/R 105/IV, Part 18.

2.1.3. Synthesis procedure

2.1.3.1. Synthesis of AB_2 Type Monomer

Diethylene triamine (0.5 mol, 52 ml) was added in a three-necked flask equipped with a constant-voltage dropping funnel, condenser and a nitrogen inlet. The flask was placed in an ice bath and solution of methyl acrylate (0.5 mol, 43 ml) in methanol (100 ml) was added dropwise into the flask. The reaction mixture was stirred with a magnetic stirrer. Then the mixture was removed from the ice bath and left to react with a flow of nitrogen at room temperature (Scheme 1). After stirring for 4 hours, the nitrogen flow was removed and AB_2 type monomer was obtained.

IR ν_{max} (cm^{-1}, KBr), 3300–3500 (N–H), 2956, 2832, 1735 (C=O).

2.1.3.2. Synthesis of Hyperbranched Polymer

The obtained light yellow and viscous mixture was transferred to an eggplant-shaped flask for an automatic rotary vacuum evaporator to remove the methanol under low pressure. Then the temperature was raised to 150 °C using an oil bath and condensation reaction was carried out for 6 hours. A pale yellow viscous hyperbranched polymer was obtained. The ^{1}H and ^{13}C-NMR spectrum was in accordance with the proposed structure.

IR ν_{max} (cm^{-1}, KBr), 3300–3500 (N–H), 1631 (C=O). ^{1}H NMR (ppm, DMSO-d_6), 4.95 (s), 3.45 (m), 2.64 (m). ^{13}C NMR (ppm, DMSO-d_6), 171.98, 59.86, 57.71, 40.68–39.68 (group).

2.1.4. PET Fabric Treatment with a Hyperbranched Polymer

PET fabrics were immersed in sodium hydroxide solution (10% w/v) at 94°C for 1 hour, using a liquor ratio of 40:1. After the alkaline treatment of fabrics, the samples were thoroughly rinsed with distilled water and neutralized with acetic acid, then rinsed and dried at room temperature. Application of a hyperbranched polymer to original PET fabric and alkali-treated PET fabric (HPET) was carried out using exhaust method. PET samples were immersed in a hyperbranched polymer solution (20 g/dL) and the temperature was raised at a rate of 2.5°C/min. After 60 min, the samples were thoroughly rinsed with distilled water to remove unfixed hyperbranched polymer and dried at room temperature. Figure 2 shows this application profile.

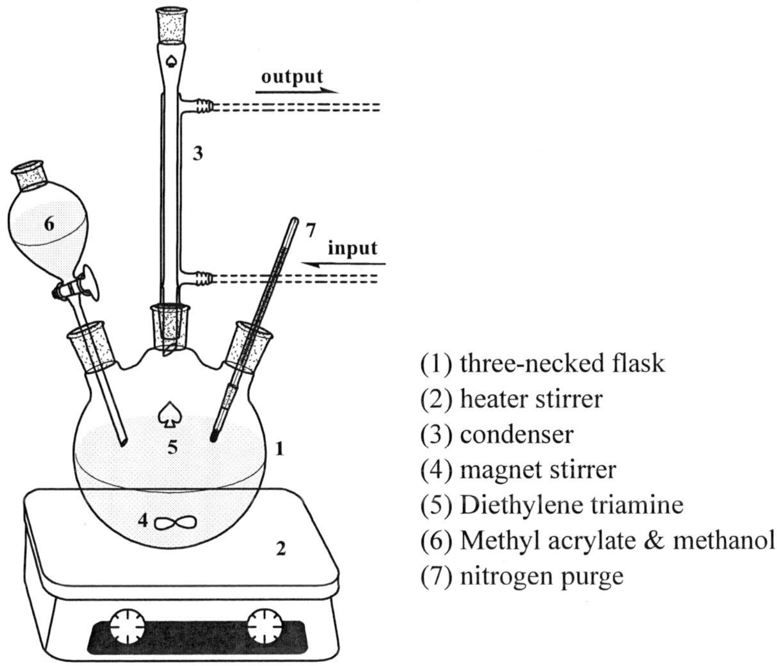

Scheme 1. Schematic of reactor for monomer synthesis.

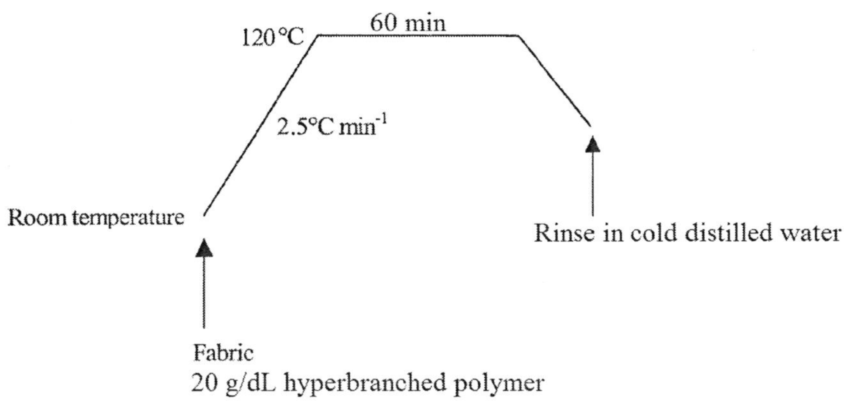

Figure 2. Hyperbranched polymer application profile.

The HBP-treated and untreated PET fabrics were subjected to five repeated washing fastness test for 30 min using sodium carbonate (2 g/L) and detergent (5 g/L) at 60°C, as shown in Figure 3.

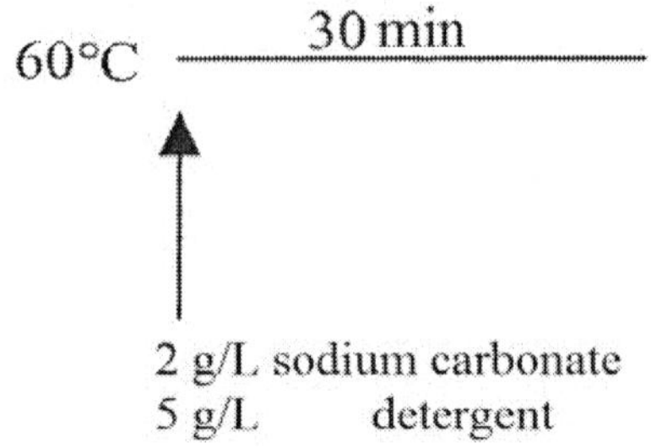

Figure 3. Washing fastness test profile.

2.1.5. Dyeing Procedure with Acid Dye

The dyeability of the hyperbranched-treated PET fabric and untreated polyester fabric were examined using C.I. Acid Red 114 (AR 114). All samples were dyed competitively in the same dye bath. The treated PET fabrics were introduced into the dye bath at 40 °C, then the temperature was gradually raised up to boiling point within 20 min, and then dyeing was continued for 60 min with occasional stirring. The dyeing process was carried out using a liquor ratio of 60:1. At the end of dyeing, the dyed samples were rinsed with cold water, then with hot water at about 50 °C and finally rinsed with tap water. Figure 4 shows the dyeing profile.

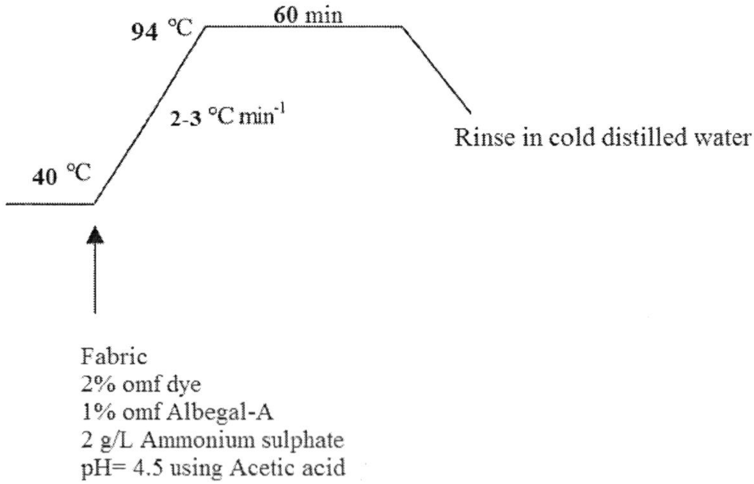

Figure 4. Dyeing method.

2.2. Case 2- Experimental Design and Optimization f Treatment Conditions

Response surface methodology (RSM) is a combination of mathematical and statistical techniques used to evaluate the relationship between a set of controllable experimental factors and observed results. This optimization process involves three major steps: (i) performing statistically designed experiments, (ii) estimating the coefficients in a mathematical model, and (iii) predicting the response and checking the adequacy of the model [36]. RSM is used in situations where several input variables influence some output variables (responses) of the system. The main goal of RSM is to optimize the response, which is influenced by several independent variables, with a minimum number of experiments. RSM, which is a powerful and useful experimental design tool, is capable of fitting a second-order (quadratic) prediction equation for the response. Central Composite Design (CCD), first introduced by Box and Wilson, is the most common type of second-order designs that used in RSM and is appropriate for fitting a quadratic surface [37].

In the present study, CCD was employed for the optimization of PET fabrics treated with hyperbranched polymer as shown in Table 1. The experiment was performed for at least three levels of each factor to fit a quadratic model. Based on preliminary experiments, hyperbranched polymer solution concentration (X_1), treatment temperature (X_2), and time (X_3) were determined as critical factors with significance effect on treated PET fabrics. These factors were three independent variables and chosen equally spaced, while the fabrics dyeability (K/S value) was dependent variable (response). A schematic of experimental design is shown in Figure 5. -1, 0, and 1 are coded variables corresponding to low, intermediate and high levels of each factor, respectively. The actual design experiment for three independent variables listed in Table 2.

Table 1. Design of experiment (factors and levels)

Factor	Variable	Unit	Factor level		
			-1	0	1
X_1	Hyperbranched polymer concentration	(wt.%)	2	6	10
X_2	Temperature	(°C)	90	110	130
X_3	Time	(min)	45	60	75

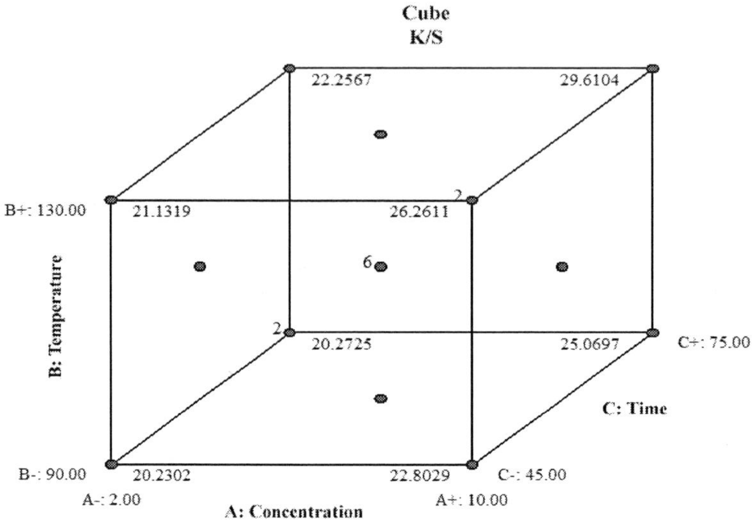

Figure 5. Design of experiment schematic.

The following quadratic model, which also includes the linear model, was fitted to the data presented in Table 2:

$$Y = \beta_0 + \sum_{i=1}^{k} \beta_i . x_i + \sum_{i=1}^{k} \beta_{ii} . x_i^2 + \sum \sum_{i<j=2}^{k} \beta_{ij} . x_i x_j + \varepsilon \qquad (2)$$

where, Y is the predicted response, x_i and x_j are coded variables, β_0 is constant coefficient, β_i is the linear coefficients, β_{ii} is the quadratic coefficients, β_{ij} is the second-order interaction coefficients, k is the number of factors, and ε is the approximation error.

The experimental data in Table 2 was analyzed using Design-Expert® software including analysis of variance (ANOVA). The values of coefficients for parameters (β_0, β_i, β_{ii}, β_{ij}) in equation 2 and p-values are calculated by regression analysis. The coefficient of determination (R^2) has been explained by the regression model.

Table 2. Design of experiment (coded values)

Expt. no.	Coded value		
	X_1	X_2	X_3
1	-1	-1	1
2	-1	1	1
3	-1	0	0
4	-1	1	-1
5	-1	-1	-1
6	0	0	0
7	0	0	0
8	0	0	0
9	0	0	0
10	0	0	0
11	0	0	0
12	0	0	-1
13	0	0	1
14	0	-1	0
15	0	1	0
16	1	-1	-1
17	1	-1	1
18	1	1	1
19	1	1	-1
20	1	0	0

3. Results and Discussion

3.1. Synthesis of Hyperbranched Polymer

As mentioned earlier, preparation of hyperbranched polymer requires two-step reaction comprise preparation of AB_2 type monomers (1 and 2) by Michael addition reaction of methyl acrylate and diethylene triamine and synthesis of hyperbranched polymer by polycondensation reaction, respectively (Scheme 2). It is well known that in hyperbranched polymers, the structural units include terminal (T), dendritic (D) and linear (L) units that are shown in Scheme 2.

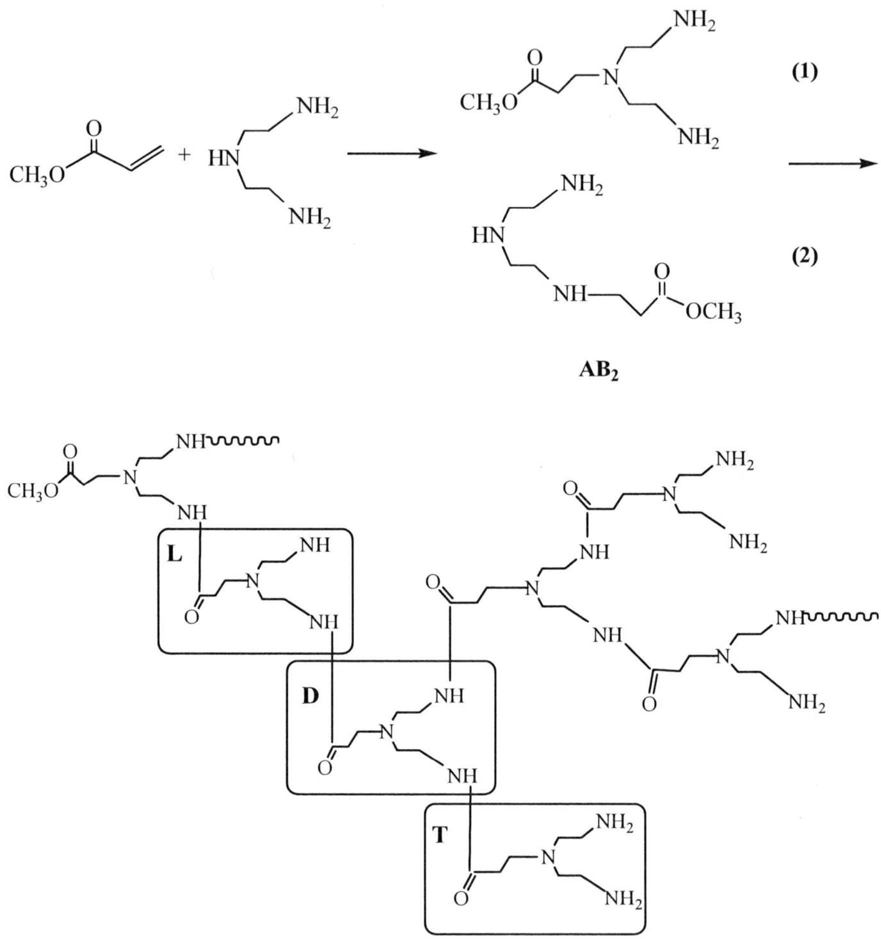

Scheme 2. Chemical structure of amine-terminated hyperbranched polymer.

3.2. FTIR Spectroscopy Characterization

3.2.1. Hyperbranched Polymer Analysis

To confirm the polycondensation reaction progress, AB_2 type monomer and hyperbranched polymer were examined by FTIR spectroscopy. Spectral differences between AB_2 type monomer and hyperbranched polymer are observed in the fingerprint region between 1800 and 900 cm^{-1}. The absorption bands in FTIR spectra are assigned according to literature [38]. FTIR spectra of the AB_2 type monomer and the hyperbranched polymer are given in Figure 6. The FTIR spectra of both monomer and polymer show the peak position at above 3300 cm^{-1} due to N–H stretching vibration. AB_2 type monomer had absorption band at around 1463 cm^{-1} (bending vibration of CH_2) and bands between 1000 and 1400 cm^{-1}. The peaks can be observed at 1735 cm^{-1} corresponding to C=O stretching vibration of esters.

FTIR spectrum of hyperbranched polymer is characterized by absorption at around 1631 cm^{-1} (C=O stretching vibration of amides) and 1463 cm^{-1} (bending vibration of CH_2). The absorption at 1735 cm^{-1} that is attributed to the C=O stretching vibration of esters is generally weak in the hyperbranched polymer. Melt-polycondensation reaction and synthesis can be responsible for this change in the band intensity.

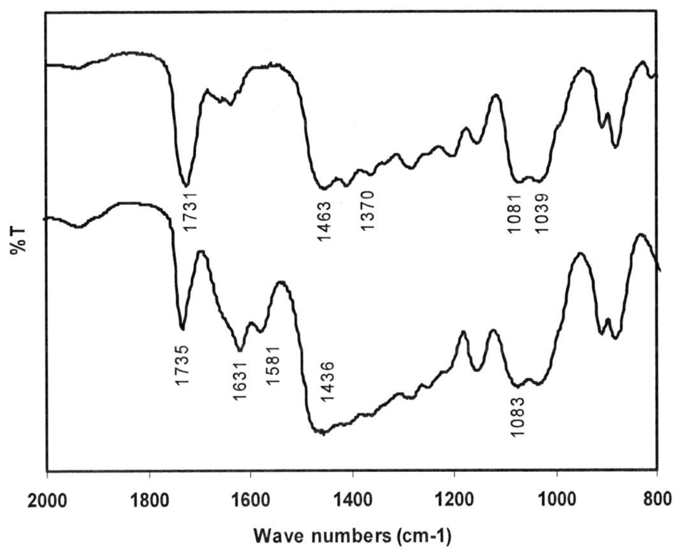

Figure 6. FTIR spectra of (a) AB_2-type monomer and (b) hyperbranched polymer.

3.2.2. Hyperbranched Treated PET Fabrics Analysis

In order to confirm the preparation process, we studied the FTIR spectroscopic data. FTIR spectra of untreated PET fabric and hyperbranched polymer-treated PET fabric are shown in Figure 7. The FTIR of both samples shows that the peak positions are at 2929, 2867, 1714, 1238, and 1093 cm^{-1}. The band at 1238, and 1093 cm^{-1} is due to C-O stretching, while the band at

1714 cm^{-1} reflects the carbonyl group stretching. The characteristics absorption band of hyperbranced polymer, around 1600 cm^{-1}, which reflects the N-H bending of hyperbranched polymer, appeared after the treatment of PET fabric.

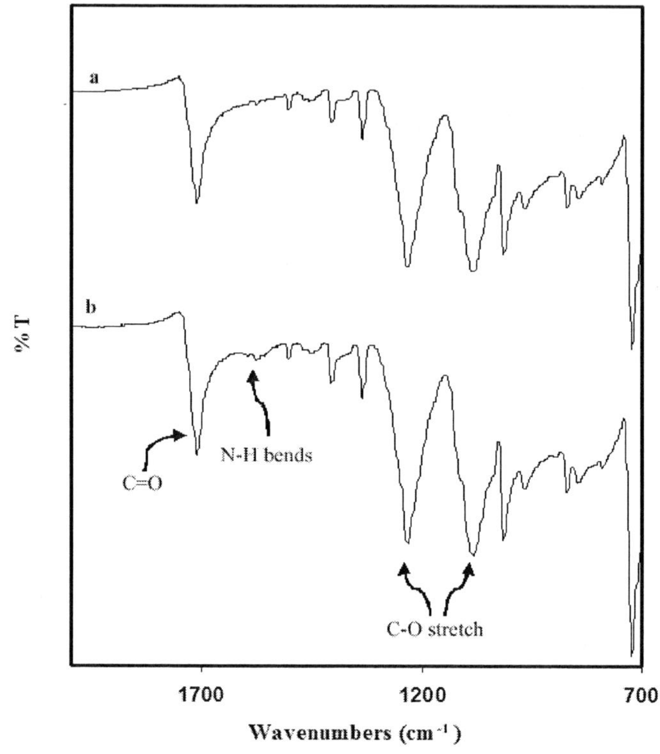

Figure 7. FTIR spectra of (a) untreated PET fabric and (b) HBP-treated PET fabric.

3.3. NMR Spectroscopy Characterization

3.3.1. ^{1}H NMR Analysis

The possible structure of the novel hyperbranched polymer was supported with NMR spectroscopy. Figure 8 shows the ^{1}H NMR spectrum and its assignment of amine-terminated hyperbranched polymer. A good correlation is found between the ^{1}H NMR spectrum of hyperbranched polymer and its expected structure.

The chemical shifts ranged from δ = 2.80 to 3.70 ppm were assigned to protons of methylene. The methylene protons adjacent to the carbonyl unit give rise to a signal at δ = 2.90 ppm. The presence of the amine unit is confirmed by the resonance at δ = 2.74 ppm for the NH_2 of the amine group. The chemical shift at δ = 2.50 ppm are related to the DMSO-d_6 as a solvent [39, 40]. The proton absorption of amid group (-CONH-) was found between δ =5.15 and 5.40 ppm with its peak at δ =5.28 ppm [41].

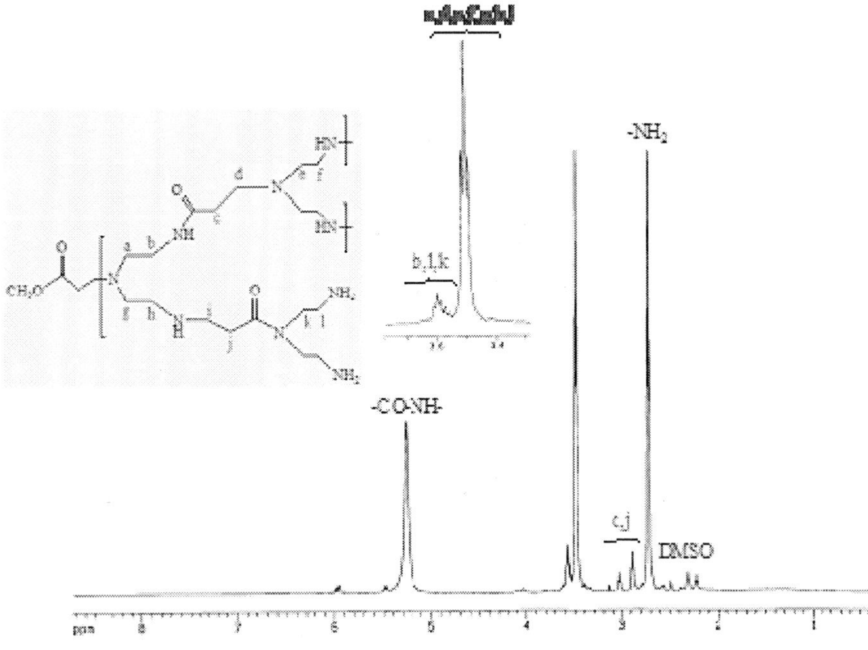

Figure 8. ^{1}H NMR spectrum of amine-terminated hyperbranched polymer.

3.2.2. ^{13}C NMR Analysis

Both ^{1}H NMR and ^{13}C NMR have been utilized to confirm the structure of hyperbranched polymer. The ^{13}C NMR spectrum of hyperbranched polymer in DMSO-d_6 is shown in Figure 9.

This spectrum is in conformity with the expected branched structure. The methylene carbon appears at δ = 50-60 ppm. The chemical shift at δ = 40 ppm are due to the solvent (DMSO) and the carbonyl carbon (-C=O) resonance is downfield at δ = 171.97 ppm.

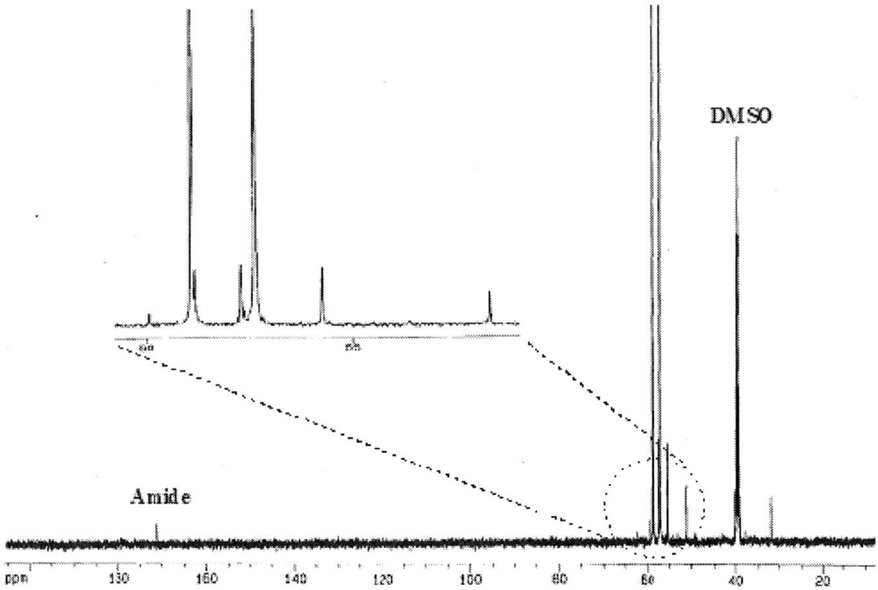

Figure 9. ^{13}C NMR spectrum of amine-terminated hyperbranched polymer.

3.3. UV-vis Spectroscopy

Absorption characteristics of hyperbranched polymer were investigated by UV–vis spectrophotometry. The UV-vis absorption spectrum of hyperbranched polymer in water solution (consentration: 20 g/dl) is shown in Figure 10. It is obvious that the synthesized hyperbranched polymer gives strong absorption in the range of 200-350 nm. Subsequently, the hyperbranched polymer treatment can improve the ultraviolet protection property of PET fabric.

3.3. Solubility Properties

Solubility properties of the hyperbranched polymer are determined in different organic solvents included polar and non-polar solvents. The hyperbranched polymer is found to be soluble in H_2O, DMSO and DMF and insoluble in THF, xylene and toluene. Because of its highly branched structure and the nature of end functional groups, this polymer was highly soluble in polar solvents such as H_2O, DMSO and DMF, as shown in Table 3.

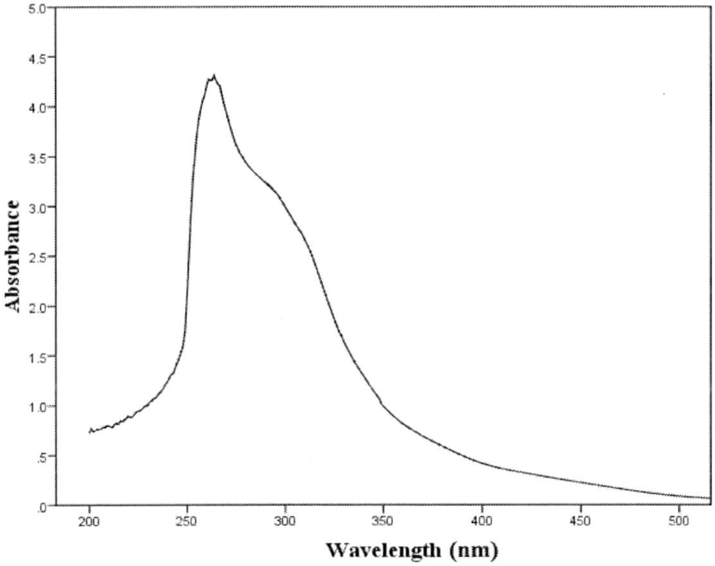

Figure 10. UV-vis spectrum of hyperbranched polymer in water.

Table 3. Solubility[*] properties of hyperbranched polymer

Polymer	Solvent						
	H_2O	DMSO	DMF	THF	Toluene	Xylene	Ethanol
Hyperbranched polymer	+	+	+	−	−	−	±

[*] The amount of polymer sample and solvents are 20 mg and 2 ml, + : soluble at room temperature (25 °C), ± : partially soluble, − : insoluble at room temperature.

3.4. Contact Angle Measurement

3.4.1. HBP Contact Angle

Contact angle measurements were carried out by experimental equipment consisting of a camera, computer, and monitor. The contact angle of hyperbranched polymer aqueous solution droplet placed on a polyester plate is shown in Figure 11. Comparison of the droplet by contact angle measurements shows that the contact angle decreased dramatically with increasing concentration of hyperbranched polymer solution from 2.5 g/dL to 20 g/dL.

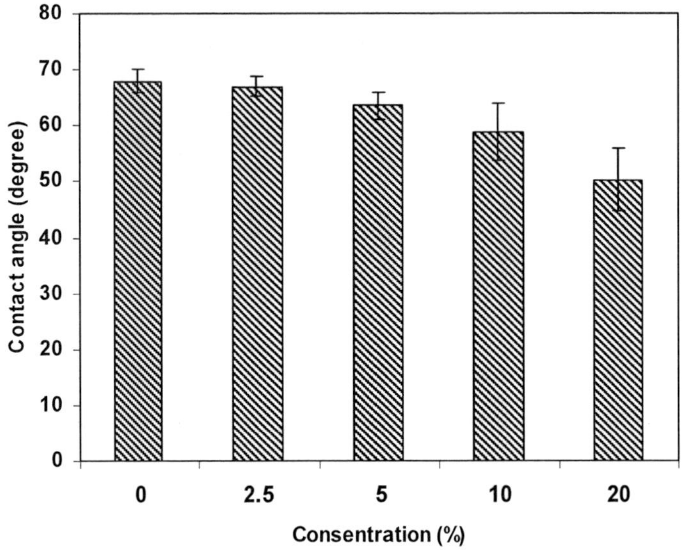

Figure 11. Contact angle of hyperbranched polymer solutions on PET plate.

3.4.1. HBP-Treated PET Fabrics Contact Angle

Table 4 shows the corresponding values of the contact angle untreated and HBP-treated PET fabrics. It is obvious that the alkali treatment considerably influenced the contact angle of PET fabrics and when the HPET fabric was treated with HBP, the contact angle was not changed significantly. But when the PET fabric was treated with HBP (PET-HBP), the contact angle was considerably decreased and the HBP-treated fabrics become hydrophilic. This confirms the presence of the HBP on the PET fabrics surface, which remains more hydrophilic than the untreated PET fabric.

Complete wetting of fiber surfaces is critical in all wet processes for textiles, particularly for dyeing. Uneven fiber wetting in dyeing invariably leads to uneven dye absorption. Therefore, HBP treatment causes PET fabric surfaces to become hydrophilic and dyeing properties of PET fabrics with acid dyes become modified.

Table 4. Contact angle of untreated and HBP-treated PET fabrics

No.	Sample	Contact angle (°)	
		Advancing	Receding
1	PET	88 ±6	85 ±4
2	HPET	56 ±6	50 ±8
3	PET-HBP	58 ±6	55 ±4
4	HPET-HBP	61 ±6	59 ±6

3.5. X-Ray Diffraction (XRD)

Figure 12 illustrate wide-angle X-ray patterns of polyester samples including untreated PET, HPET and HBP-treated fabrics. All of the samples have characteristic peaks around 2θ=17, 22 and 25° corresponding to plans with Miller indices (010), (110) and (100), which are in good agreement with other papers [42]. As can be seen from Figure 12, the HBP-treated PET fabrics exhibit a similar XRD pattern to untreated PET fabric, but the intensity of the 2θ =15° to 30° peak decreased for the PET-HBP sample, which suggested a loss of crystalline order with applying hyperbranched polymer.

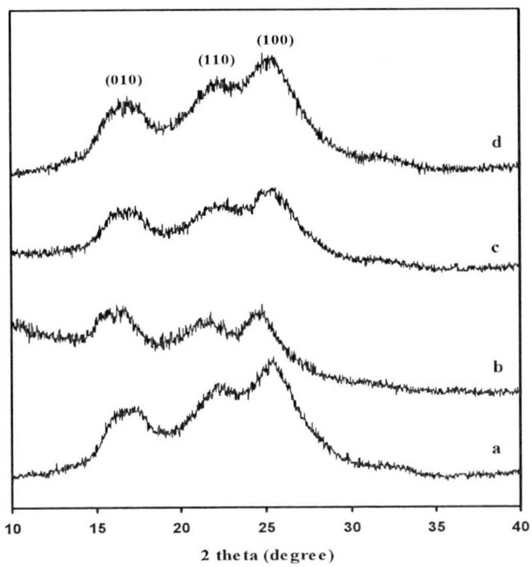

Figure 12. XRD patterns of (a) PET, (b) HPET, (c) PET-HBP and (d) HPET-HBP fabrics.

The d-sapacing of crystalline part of the polymer was calculated using the Bragg equation:

$$d = \frac{\lambda}{2\sin\theta} \tag{3}$$

where d is Bragg spacing, θ is the diffraction angle and λ is the wavelength (1.54 Å) of X-ray radiation used.

A change in the crystallite size was observed after HBP treatment. The Scherrer formula [42] was used to calculate the crystallite size as given by

$$l_{(hkl)} = \frac{k\lambda}{\Delta_{(hkl)}\cos\theta} \tag{4}$$

where $l_{(hkl)}$ is the dimension of crystallite size, Δ is the full width at half maximum intensity (FWHM) of the diffraction peak and k is a constant assigned as a 1. Table 5 shows the XRD parameters for untreated and HBP treated PET fabrics.

Table 5. XRD parameters for untreated and HBP-treated PET fabrics

No.	Sample	Reflection index	*hkl*		
			010	110	100
1	PET	Crystallite size (Å)	30.9	33.6	25.8
		2θ (deg.)	17.22	22.58	25.48
		Bragg spacing (Å)	5.14	3.93	3.49
2	HPET	Crystallite size (Å)	25.6	33.6	30.5
		2θ (deg.)	16.25	21.57	24.81
		Bragg spacing (Å)	5.45	4.11	3.59
3	PET-HBP	Crystallite size (Å)	25.3	29.8	25.3
		2θ (deg.)	16.83	22.41	25.48
		Bragg spacing (Å)	5.26	3.96	3.49
4	HPET-HBP	Crystallite size (Å)	26.0	33.9	25.4
		2θ (deg.)	16.86	22.28	25.39
		Bragg spacing (Å)	5.25	3.99	3.50

Table 5 shows that the d-spacing of crystalline part of HBP-treated samples are higher than untreated PET fabrics.

The broadening in the peak and change in the crystallite size due to treatment HBP to PET fabrics may be due to the formation of a disordered system in the HBP-treated fabric.

3.6. Zeta Potential

The surface charge of samples has been investigated by zeta potential measurement to understand how HBP act on the PET fabrics. The zeta potential values of the untreated (HPET) and the HBP-treated PET fabrics (HPET-HBP) as a function of the pH of the liquid phase are shown in Figure 13. From the results, it can be seen that the zeta potentials for the untreated PET fabrics over the entire pH range are negative, while the HBP-treated PET fabrics exhibited a positive charge on the surface at a low pH. This is probably attributed to the presence of numerous terminal primary amino groups on HBP-treated PET fabrics, that will protonate in the liquid phase and give rise to positive charge at lower pH values. It implies that HBP enhances the adsorption of acid dye on PET fabrics.

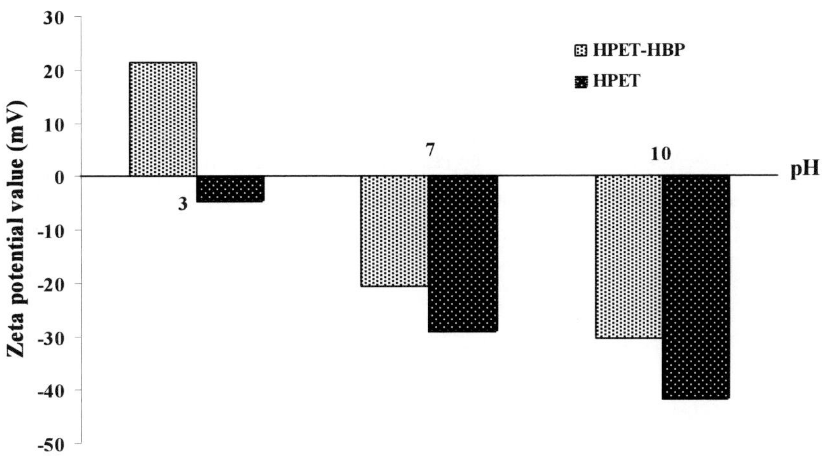

Figure 13. Zeta potential values of the untreated (HPET) and HBP treated PET fabrics (HPET-HBP).

3.7. Dyeing Properties of HBP Treated Polyester Fabric

The effect of HBP treatment on dye absorbance behavior (dyeability) of PET fabrics is evaluated by measuring its optical properties. The reflectance spectra of HBP-treated and untreated PET fabrics are shown in Figure 14 for dyed samples. As shown in Figure 14, the reflectance spectrum of HBP treated PET fabric is less than other sample. Subsequently, the treatment of polyester fabrics by HBP significantly changes the reflectance of samples.

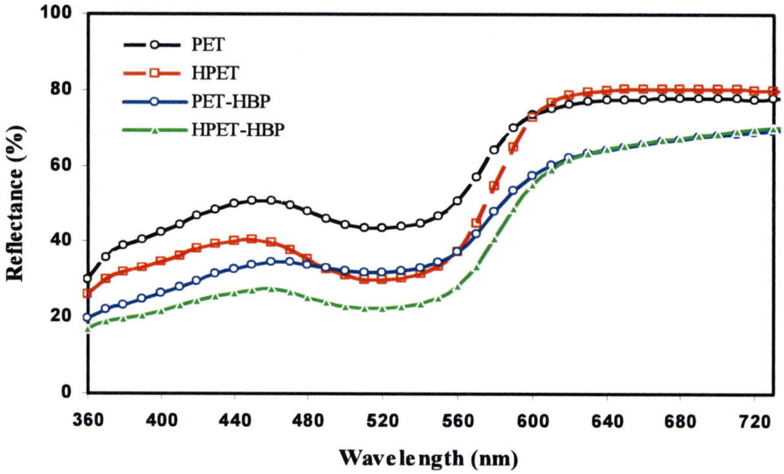

Figure 14. Reflectance spectra of dyed HBP treated and untreated PET fabrics.

The CIELAB color parameter (L*, a*, b*, C* and h°) of HBP-treated and untreated PET fabrics are shown in Table 6. As shown in this table, the lightness (L*) of untreated polyester fabric is more than other sample. Also, the hue (h°) of HBP-treated PET fabric is more than untreated sample. Subsequently, the HBP treatment has significant change on color parameters of PET fabrics.

Table 6. Color parameters of dyed samples at 2% omf acid dye

No.	Samples	L*	a*	b*	C*	h°
1	PET	78.951	18.079	5.954	19.034	0.318
2	HPET	72.780	30.075	7.577	31.015	0.247
3	PET-HBP	70.291	18.432	10.525	21.225	0.519
4	HPET-HBP	64.861	27.547	11.558	29.873	0.397

The color strength (K/S) results of the HBP-treated PET fabric and untreated PET fabric dyed with 2% omf Acid Red 114 using a competitive dyeing method, are shown in Figure 15. Higher K/S values indicate greater dye uptake and higher color strengths. It is clear that the color strength of the HBP-treated PET fabric was higher than that of the corresponding untreated PET fabric. The increase in the color strength may be because of the positively charged amino groups in the HBP treated PET fabric.

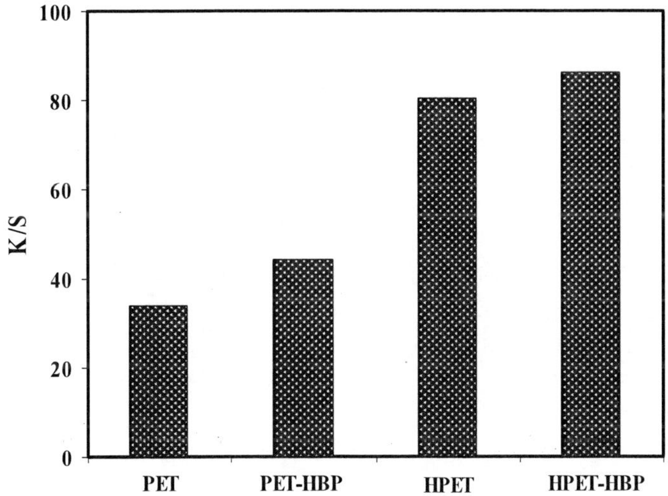

Figure 15. Effect of fabric treatment on color strength achieved using 2% omf C.I. Acid Red 114.

3.8. Fastness Properties

3.8.1. HBP Treatment Fastness

The treated samples were subjected to five repeated washing fastness test at 60 °C. Figure 16 shows the washing fastness result of the undyed HBP treated and untreated PET fabric and its effect on dyeability of PET fabrics. It is obvious that the extent of color strength change was relatively small for treated and untreated PET fabrics.

3.8.2. Dyeing Fastness

Table 7 shows the washing fastness and rubbing fastness of untreated PET fabrics (PET), hydrolyzed PET fabrics (HPET) and HBP treated PET fabrics (PET-HBP and HPET-HBP) dyed with acid dye. Fastness results show that the rubbing fastness of treated samples were good; the washing fastness of all sample were, however, poor and unaffected by treatment with HBP.

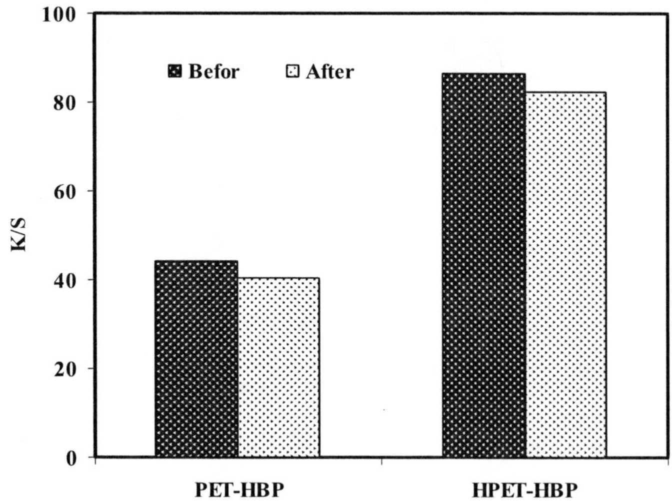

Figure 16. Effect of washing fastness test on dyeability of treated and untreated PET fabrics.

Table 7. Fastness properties of HBP treated and untreated PET fabrics

No.	Samples	Washing fastness*	Rubbing fastness**	
			Dry	Wet
1	PET	1	4	3-4
2	HPET	1	5	4-5
3	PET-HBP	1	4-5	4
4	HPET-HBP	1	4-5	4

* With gray scale.

** With stain scale.

3.9. The Analysis of Variance (ANOVA) and Optimization

All 20 experimental runs of central composite design (CCD) were performed in accordance with Table 8.

Table 8. The actual design of experiments and response

Un-coded value			Response
HBP concentration, (wt.%)	Temperature, (°C)	Time, (min)	K/S value
2	90	75	20.38
2	130	75	21.57
2	110	60	19.86
2	130	45	21.09
2	90	45	19.78
6	110	60	22.68
6	110	60	22.18
6	110	60	22.67
6	110	60	22.97
6	110	60	22.64
6	110	60	22.26
6	110	45	22.59
6	110	75	24.47
6	90	60	22.58
6	130	60	26.07
10	90	45	23.35
10	90	75	24.97
10	130	75	29.92
10	130	45	26.01
10	110	60	23.23

A significance level of 5% was selected; that is, statistical conclusions may be assessed with 95% confidence. In this significance level, the factor has a significant impact on fabric's dyeability if the p-value is less than 0.05. And when p-value is greater than 0.05, it is concluded the factor has no significant impact on the fabric's dyeability. The results of analysis of variance (ANOVA) are shown in Table 9. The regression equation was obtained from the analysis of variance.

Table 9. Analysis of variance for response surface

p-value	F-value	Sum of squares	Source
< 0.0001	25.98	104.71	Model
< 0.0001	137.49	61.58	X_1 -concentration
< 0.0001	41.33	18.51	X_2 - temperature
0.0025	16.05	7.19	X_3 -time
0.0223	7.30	3.27	X_1X_2
0.0406	5.52	2.48	X_1X_3
0.2794	1.31	0.59	X_2X_3
0.0095	10.24	4.59	X_1^2
0.0042	13.62	6.10	X_2^2
0.1177	2.93	1.31	X_3^2
Not significant	31.17	3.41	Lack of Fit
	Adj- R^2 : 0.922	R^2 : 0.959	Std.dev : 0.67

$$K/S = +22.57 + 2.48X_1 + 1.36X_2 + 0.85X_3 + 0.64X_1X_2 + 0.56X_1X_3 + 0.27X_2X_3 - 1.29X_1^2 + 1.49X_2^2 + 0.69X_3^2 \quad (5)$$

From the p-values presented in Table 3, it can be concluded that P-value of term X_3 is greater than the P-values for terms X_1 and X_2. Also P-value of term X_3^2 and X_2X_3 is much greater than the significance level of 0.05. And the interaction between temperature and time is not significant. But P-values for terms related to X_1 and X_2 are less than 0.05. Therefore, HBP concentration and treatment temperature have a significant impact on the HBP treated fabrics dyeability.

Because terms related to treatment time have no significant impact on samples K/S values, we removed term X_3^2 and X_2X_3 and fitted the equation by regression analysis again. The fitted equation in the coded unit is given by

$$K/S = +22.65 + 2.48X_1 + 1.36X_2 + 0.85X_3 + 0.64X_1X_2 + 0.56X_1X_3 - 1.03X_1^2 + 1.75X_2^2 \quad (6)$$

Now, all of the P-values are less than the significance level of 0.05.

The effects of three variables on K/S value of dyed samples are shown in Figure 17. Figure 17 (a) shows K/S value of dyed samples at different HBP concentration and temperature for constant treatment time. The surface plot shows that at any given temperature the K/S value of samples increase with increasing the HBP concentration. Maximum K/S value was observed at high HBP concentration and high temperature. Figure 17 (b) shows the response surface plot of interactions between HBP concentration and treatment time. The increase in K/S value with an increase in concentration at any given time can be seen. Moreover, it should be noted that at higher temperatures, the increase in concentration and time results in the higher K/S value.

The actual and the predicted K/S value of samples plot with correlation coefficient of 0.91, is shown in Figure 18. Actual values are the measured response data for a particular run and the predicted values evaluated from the model. It can be observed that experimental values are in good agreement with the predicted values. The optimal conditions to obtain the maximum dyeability (K/S value) efficiency are shown in Table 10.

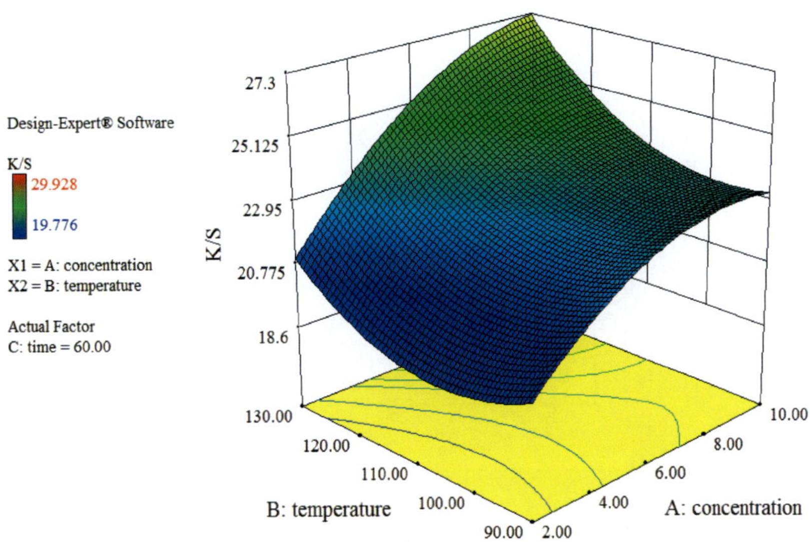

Figure 17. (Continued).

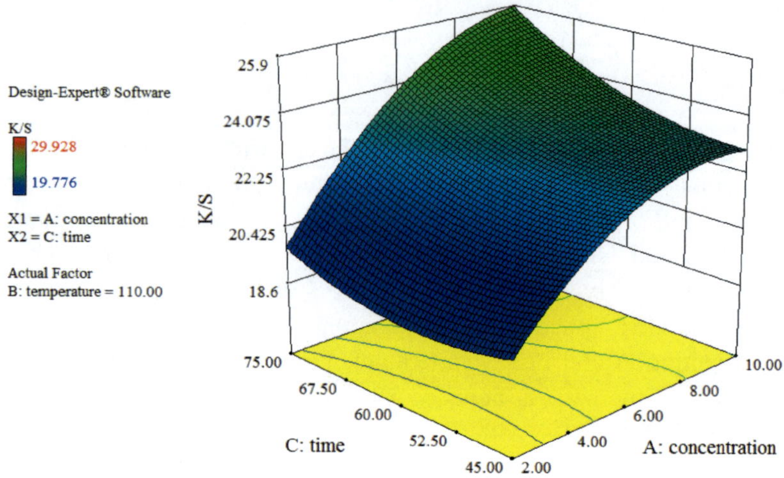

Figure 17. Response surface for K/S value of dyed samples in terms of: (a) HBP concentration and treatment temperature, (b) HBP concentration and treatment time.

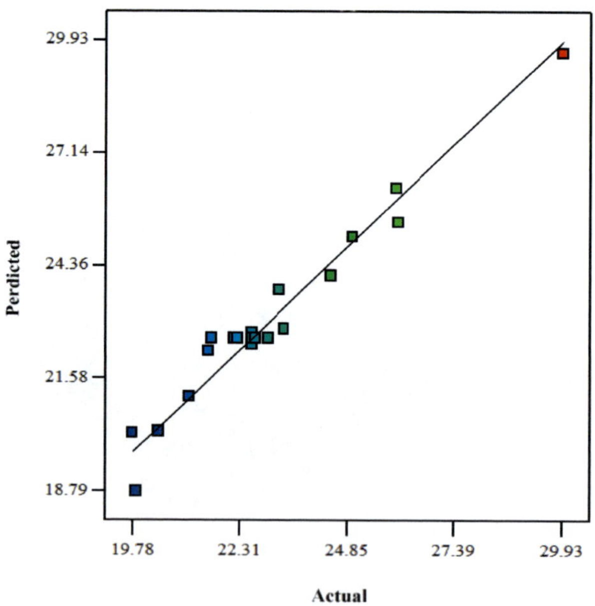

Figure 18. The actual and predicted plot for K/S value of dyed samples.

Table 10. Optimum values of the treatment conditions for maximum dyeability efficiency

HBP concentration (wt.%)	Temperature (°C)	Time (min)	Predicted K/S	Experimental K/S	Desirability
10	130	75	29.25	29.93	0.93

CONCLUSION

In this study, a hyperbranched, functional and water-soluble polymer was successfully synthesized by melt polycondensation reaction of methyl acrylate and diethylene triamine. The FTIR, ^{1}H and ^{13}C NMR spectroscopy measurements of hyperbranched polymer indicated that this polymer comprised terminal amine group and branching occurred. The solubility properties of resulting hyperbranched polymer indicated that this polymer is readily soluble in polar solvents such as water, DMSO and DMF. The hyperbranched polymer was applied to alkali-treated PET fabric using the exhaust method. Using Bragg's equation, it was found that d-spacing increased after applying HBP to PET fabrics. The study of dyeability of treated samples with C.I. Acid Red 114 indicated that the color strength of HBP-treated PET fabrics is more than untreated PET fabric due to the presence of terminal primary amino groups in the molecular structure of the HBP, that will protonate in the liquid phase and give rise to a positive charge at lower pH values, as shown by zeta potential measurement. Samples contact angles tends to decrease with the HBP treatment, and the hydrophobic, untreated PET fabric's surface becomes hydrophilic. This confirms the presence of the HBP on the PET fabric's surface and its effect on hydrophilic behavior of PET fabrics. It is noteworthy that in our attempt, PET fabric dyed with acid dye effectively using HBP. Compared with a previously reported article, our attempt has an innovation in PET dyeing with acid dyes.

The concentration, temperature and time are the most important factors for HBP treatment to PET fabrics and dyeability efficiency. Response surface methodology was successfully applied to find out the optimum level of the above factors using central composite design (CCD). The optimum concentration of HBP, temperature and time were found to be 10 wt.%, 130°C and 75 min, respectively, for a maximum K/S value of HBP treated samples.

References

[1] Jikei, M., Kakimoto, M. Hyperbranched polymers: a promising new class of materials. *Progress in Polymer Science*. 2001; 26: 1233-1285.

[2] Gao, C., Yan, D. Hyperbranched polymers: from synthesis to applications. *Progress in Polymer Science*. 2004; 29: 183-275.

[3] Voit, B.I., Lederer, A. Hyperbranched and highly branched polymer architectures synthetic strategies and major characterization aspects. *Chem. Rev.* 2009; 109: 5924–5973.

[4] Kumar, A., Meijer, E.W. Novel hyperbranched polymer based on urea linkages. *Chem. Commun.* 1998; 1629-1630.

[5] Grabchev, I., Petkov, C., Bojinov, V. Infrared spectral characterization of poly(amidoamine) dendrimers peripherally modified with 1,8-naphthalimides. *Dyes and Pigments*. 2004; 62: 229–234.

[6] Qing-Hua, C., Rong-Guo, C., Li-Ren, X., Qing-Rong, Q., Wen-Gong, Z. Hyperbranched poly(amide-ester) mildly synthesized and its characterization. *Chinese J. Struct. Chem.* 2008; 27: 877-883.

[7] Kou, Y., Wan, A., Tong, S., Wang, L., Tang, J. Preparation, characterization and modification of hyperbranched polyester-amide with core molecules. *Reactive & Functional Polymers*. 2007; 67: 955–965.

[8] Schmaljohann, D., Pötschke, P., Hässler, R., Voit, B.I., Froehling, P.E., Mostert, B., Loontjens, J.A. Blends of amphiphilic, hyperbranched polyesters and different polyolefins. *Macromolecules*. 1999; 32: 6333-6339.

[9] Kim, Y.H. Hyperbranched polymers 10 years after. *Journal of Polymer Science: Part A: Polymer Chemistry*. 1998; 36: 1685–1698.

[10] Liu, G., Zhao, M. Non-isothermal crystallization kinetics of AB3 hyper-branched polymer/polypropylene blends. *Iranian Polymer Journal*. 2009; 18: 329-338.

[11] Seiler, M. Hyperbranched polymers: Phase behavior and new applications in the field of chemical engineering. *Fluid Phase Equilibria*. 2006; 241: 155-174.

[12] Yates, C.R., Hayes, W. Synthesis and applications of hyperbranched polymers. *European Polymer Journal*. 2004; 40: 1257-1281.

[13] Voit, B. New developments in hyperbranched polymers. *Journal of Polymer Science: Part A: Polymer Chemistry*. 2000; 38: 2505-2525.

[14] Nasar, A.S., Jikei, M., Kakimoto, M. Synthesis and properties of polyurethane elastomers crosslinked with amine-terminated AB2-type

hyperbranched polyamides. *European Polymer Journal.* 2003; 39: 1201-1208.

[15] Froehling, P.E. Dendrimers and dyes-a review. *Dyes and pigments.* 2001; 48: 187-195.

[16] Jikei, M., Fujii, K., Kakimoto, M. Synthesis and characterization of hyperbranched aromatic polyamide copolymers prepared from AB2 and AB monomers. *Macromol. Symp.* 2003; 199: 223-232.

[17] Radke, W., Litvinenko, G., Müller, A.H.E. Effect of core-forming molecules on molecular weight distribution and degree of branching in the synthesis of hyperbranched polymers. *Macromolecules.* 1998; 31: 239-248.

[18] Maier, G., Zech, C., Voit, B., Komber, H. An approach to hyperbranched polymers with a degree of branching of 100%. *Macromol. Chem. Phys.* 1998; 199: 2655–2664.

[19] Voit, B., Beyerlein, D., Eichhorn, K., Grundke, K., Schmaljohann, D., Loontjens, T. Functional hyper-branched polyesters for application in blends, coations, and thin films. *Chem. Eng. Technol.* 2002; 25: 704-707.

[20] Gao, C., Xu, Y., Yan, D., Chen, W. Water-Soluble degradable hyperbranched polyesters: novel candidates for drug delivery?. *Biomacromolecules.* 2003; 4: 704-712.

[21] Paleos, C.M., Tsiourvas, D., Sideratou, Z. Molecular engineering of dendritic polymers and their application as drug and gene delivery systems. *Molecular pharmaceutics.* 2007; 4:169-188.

[22] Wu, D., Liu, Y., Jiang ,X., He, C., Goh, S.H., Leong, K.W. Hyperbranched poly(amino ester)s with different terminal amine groups for DNA delivery. *Biomacromolecules.* 2006; 7: 1879-1883.

[23] van Benthem, Rolf A.T.M. Novel hyperbranched resins for coating applications. *Progress in Organic Coatings.* 2000; 40: 203-214.

[24] Tang, W., Huang, Y., Meng, W., Qing, F. Synthesis of fluorinated hyperbranched polymers capable as highly hydrophobic and oleophobic coating materials. *European Polymer Journal.* 2010; 46: 506–518.

[25] Zhang, W., Zhang, Y., Chen, Y. Modified brittle poly(lactic acid) by biodegradable hyperbranched poly(ester amide). *Iranian Polymer Journal.* 2008; 17: 891-898.

[26] Astruc, D., Chardac, F. Dendritic catalysts and dendrimers in catalysis. *Chem. Rev.* 2001; 101: 2991-3023.

[27] Fang, J., Kita, H., Okamoto, K. Hyperbranched polyimides for gas separation applications. 1.Synthesis and Characterization. *Macromolecules.* 2000; 33: 4639-4646.

[28] Liu, C., Gao, C., Yan, D. Synergistic supramolecular encapsulation of amphiphilic hyperbranched polymer to dyes. *Macromolecules.* 2006; 39: 8102-8111.

[29] Burkinshaw, S.M., Froehling, P.E., Mignanelli, M. The effect of hyperbranched polymers on the dyeing of polypropylene fibres. *Dyes and Pigments.* 2002; 53: 229–235.

[30] Zhang, F., Chen, Y., Lin, H., Lu, Y. Synthesis of an amino-terminated hyperbranched polymer and its application in reactive dyeing on cotton as a salt-free dyeing auxiliary. *Coloration Technology.* 2007; 123: 351–357.

[31] Zhang, F., Chen, Y., Lin, H., Wang, H., Zhao, B. HBP-NH2 grafted cotton fiber: Preparation and salt-free dyeing properties. *Carbohydrate Polymers.* 2008; 74: 250–256.

[32] Zhang, F., Chen, Y., Lin, H., Zhang, D.S. Performance of cotton fabric treated with an amino-terminated hyperbranched polymer. *Fibers and Polymers.* 2008: 9: 515-520.

[33] Zhang, F., Chen, Y., Ling, H., Zhang, D. Synthesis of HBP-HTC and its application to cotton fabric as an antimicrobial auxiliary. *Fibers and Polymers.* 2009; 10: 141-147.

[34] Zhang, F., Zhang, D., Chen, Y., Lin, H. The antimicrobial activity of the cotton fabric grafted with an amino-terminated hyperbranched polymer. *Cellulose.* 2009; 16:281–288.

[35] Khatibzadeh, M., Mohseni, M., Moradian, S. Compounding fibre grade polyethylene terephthalate with a hyperbranched additive and studying its dyeability with a disperse dye. *Coloration Technology.* 2010; 126: 269-274.

[36] Murugesan, K., Dhamija, A., Nam, I., Kim, Y., Chang, Y. Decolourization of reactive black 5 by laccase: Optimization by response surface methodology. *Dyes and Pigments.* 2007; 75: 176-184.

[37] Myers, R.H., Montgomery, D.C., Anderson-cook, C.M. Response surface methodology: process and product optimization using designed experiments. 3rd ed. USA: John Wiley and Sons; 2009.

[38] Pavia, D.L., Lampman, G.M., Kriz, G.S., Vyvyan, J.R. Introduction to spectroscopy. 4th ed. US: Brooks/Cole; 2001.

[39] Tang, W., Huang, Y., Meng, W., Qing, F. Synthesis of fluorinated hyperbranched polymers capable as highly hydrophobic and oleophobic coating materials. *European Polymer Journal.* 2010; 46:506-518.

[40] Reul, R., Nguyen, J., Kissel, T. Amine-modified hyperbranched polyesters as non-toxic, biodegradable gene delivery systems. *Biomaterials.* 2009; 30: 5815-5824.

[41] Pan, B., Cui, D., Gao, F., He, R. Growth of multi-amine terminated poly(amidoamine) dendrimers on the surface of carbon nanotubes. *Nanotechnology.* 2006; 17: 2483-2489.

[42] Youssefi, M., Morshed, M., Kish, M.H. Crystalline structure of poly(ethylene terephthalate) filaments. *Journal of applied polymer science.* 2007; 106: 2703-2709.

INDEX

A

B

G

H

I

J

K

L

M

N

O

P

Q

R

S

T

U

V

W

X

Y

Z